Balance in Motion

First published in the United States by Blue Imaginarium Press,
an imprint of Blue Imaginarium Publishing.

For more information, please contact the publisher at Blue Imaginarium Publishing, www.blueimaginarium.com and imagine@blueascension.com
ISBN: 978-1-960363-01-5 (softcover) v1.0

Website

Scan to visit the book's information page

BLUE IMAGINARIUM PRESS

Balance In Motion

Cultural BioMechanics Background of Aikido Formation

Kei Izawa

Blue Imaginarium Press

Contents

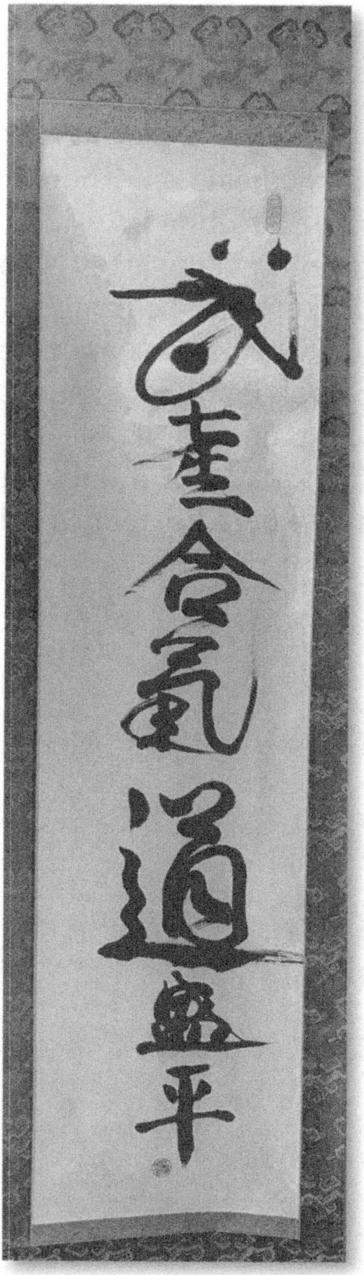

This scroll was created by Morihei Ueshiba, founder of Aikido, and the calligraphy says "Takemusu Aikido." Kanai Sensei graciously gifted us this treasure to commemorate the opening of Tanshinjuku Dojo.

Dedication

I would like to dedicate this book to Kanai Mitsunari Sensei and to my wife Mariquita Izawa. To Kanai Sensei for all the inspiration he gave me in the art of Aikido and beyond, and to my wife for all the support she has given me over the years, including many of the illustrations in this book.

The title of the book is my idea but Kanai Sensei's Aikido philosophy has much to do with it.

Let me start the book with a quote from an unpublished script of Kanai Sensei where he defines Aikido.

合気道とは

先ず身も心も智慧も練りに練りに鍛え抜く可し。

併して対決に於いては我が身を庇いて相手を害せんとする
状況を速やかに脱す可し。

その状況を我が全知全能を以て解消
せんとする意志の発動が武心（ぶごころ）である。

即ち此の武心の表現（体現）こそが合気道であると
知る可し。

武人の情（愛情）は此処より発す。

合気道師範　金井　満也

What is Aikido….

First, you must train and forge your body, mind, and wisdom to the best of your ability.
At the same time, in a confrontation, first, we must protect ourselves and quickly get out of a situation where we are trying to harm the opponent.
The activation of the will to resolve such a situation with omniscience and omnipotence is Bugokoro (true martial heart 武心).
In other words, the expression (embodiment) of this Bugokoro is Aikido.
This is where the humane feeling (love) of the warrior originates.

— Aikido Shihan Kanai Mitsunari

Thank you, Kanai Sensei, for all the teachings and inspirations.
Thank you, Mariquita, for all the support.

Kei Izawa

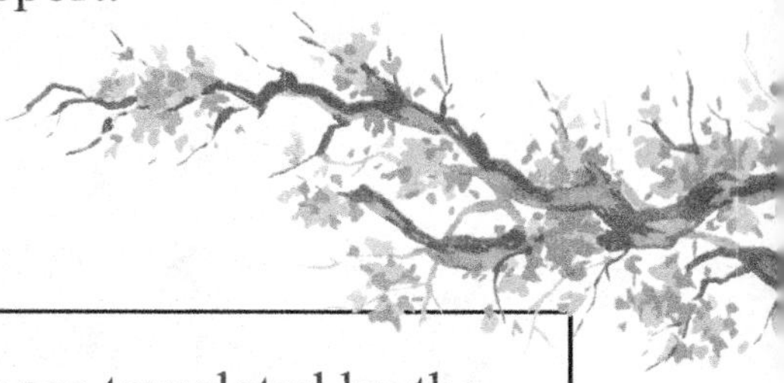

Japanese quotations in this book are translated by the author.
Where Japanese terms are used, they are shown in bold font at first reference.
Japanese terms can be found listed alphabetically in the Glossary.

01 In the Beginning

There are countless books that have been written about **Aikido** (合気道). When I decided to write a book on Aikido, my initial intent was to focus on the mechanical aspects of Aikido, as I was inspired by the teachings of many direct disciples of O Sensei, Kanai Sensei (more to follow), and other excellent instructors in Japan and around the world.

Kanai Sensei always emphasized the mechanical aspect of the techniques rather than Ki or esoteric aspects that can't be explained. His Aikido was clean and very powerful. Kanai Sensei was a man of few words and never explained in full detail, but his actual Aikido techniques demonstrated the powerful results of his mechanical Aikido approach. Because of this, I wanted to focus specifically on mechanical topics not covered in other technical books.

After writing an outline of this book, I came to realize that there were many aspects of Japanese martial arts that were also naturally and deeply ingrained into the culture and tradition of Japan. I am not trying to say that Japan is different and totally unique, but out of the social structure, lifestyle, customs, social norms, etiquette, wardrobes, eating, and sleeping habits, I came to realize that some unique Japanese body movement characteristics emerged from the combined nature of the societal way of life.

These Japan-based characteristics have evolved over the centuries, as both external and some internal ones forced Japan to change in some ways. Some of the major external change forces were the Mongol invasion of Japan in 1274 and 1281 and the introduction of firearms (**Hinawaju**, matchlock gun 火縄銃) by the Portuguese in 1543.

The introduction of Hinawaju in the battles not only changed the tactics of the battles but also changed the fortifications of the castles and Samurai armor. As the mainstream fighting style shifted from group battles with horseback riders to group foot battles and indoor battles inside castles, the demand for shorter and thicker swords increased. But even with these external impacts for change, many of the Japanese martial arts kept their core mechanical moves rather intact. I must say that the societal influence was so strong that even those external pressures brought only practical changes. The inner self was still very much of Japanese tradition.

When you look at the various martial arts of Japan, whether it is **Kyudo** (archery 弓道), **Sumo** (相撲), **Kendo** (剣道), **Jujutsu** (柔術), **Naginata** (halberd 薙刀), **Yari** (spear 槍), **Yabusame** (horseback archery 流鏑馬), and other martial arts including Aikido (合気道), many of the core teachings and mechanical movements have something very uniquely Japanese and have many similarities with traditional **Japanese Dance** (日本舞踊) and **Noh** (能).

We will go into further details later in this book, but this realization made me think that in order to explain the unique characteristics of Aikido moves, maybe it was necessary to cover these unique Japanese characteristics. Maybe there are many other cultures similar to Japan but without explaining the unique background, I felt Aikido couldn't be explained in its entirety.

Let me bring one simple example. Both the West and Japan have saws to cut wood. You could say that a saw is a saw and they are all the same. The Japanese saw is called **Nokogiri** (鋸, のこぎり). The difference is that western saws cut on the push, while Japanese saws cut on the pull stroke. I am not arguing that the Japanese method is better, and it's interesting to note that the Japanese sword cuts by the same principle. You would extend the sword forward so you can slice on the "pull stroke." The same action applies to sushi chefs slicing fish for sushi. They do not cut fish on the push but rather on the pull. I am not advocating that Japan's approach is unique in all aspects, but my hope with this book is that when you look at the combined picture of different actions or tasks, you will find the Japanese logic to be very consistent.

In the chapters that follow, I will go into rather significant detail about footwear, walking, sitting and sleeping, wardrobe, and more, along with ideas of how Japanese culture and movement influence them.

Kanai Sensei didn't connect the peculiarities of Japanese culture in the overall Aikido analysis. Kanai Sensei focused on the details of the mechanical impacts: human anatomy, the projection of the technique, the destabilization of the opponent, leveraging, and many other crucial elements that are needed for a powerful Aikido technique.

I do not know if combining this technical detail with cultural issues is correct or not, and the theme of this book is certainly of my own making. I do know that the more deeply I dug into the topics of culture and perhaps a cause-and-effect result in movement, I knew it was a path I wanted to explore and share. Perhaps having lived two-thirds of my life outside of Japan, and having a non-Japanese wife who speaks perfect Japanese and knows the customs of Japan, allowed me to be a cultural interpreter of this art.

During the Covid seclusion, our dojo continued Aikido virtual classes online. I thank my students very much for their dedication to continuing to train in virtual Aikido without a physical partner. Especially when we talked on Zoom about destabilizing "a non-present opponent," it must have been hard for the team. It would be like dancing tango alone with an imaginary partner.

Making bold steps anticipating a counterweight that doesn't exist is very hard. So instead, we focused on sections of the body and created new perspectives on how to execute the techniques. While preparing for the classes, I took some time to research the anatomy and its impact on the techniques. I also compared other Japanese martial arts as to how they would use those parts of the body.

For example, it was interesting to find out more about the concepts of **Hizanuki** (unplugging the knee 膝抜き) or the applications of O Sensei as to how he used to unlock the Uke's neck to apply many techniques, not highlighted in current Aikido instructions. Or the background of the Ogasawara Samurai etiquette for overall body movements applicable to Samurais.

After we resumed in-person training, it was a nice experience to put those theories to the test. My Tanshinjuku Dojo members have had much to contribute by experimenting with the hypothesis and theories to see if they were effective or not. I value very much their inquisitive nature to find ways to be more effective in executing the techniques.

The title of this book, *Balance in Motion*, perhaps applies to most martial arts and sports, as balance is crucial for the execution of any technique in any physical activity. To be a great art, any physical sport or activity must include the concept and execution of Balance in Motion, just as it would be included in classical ballet. A stable and balanced style of Aikido is not only beautiful but it is also very powerful. In this book, I tried to focus on Aikido, with some complementary analysis from other martial arts or sports.

While trying to write a book on Aikido from a physical/mechanical point of view highlighting the concept of Balance in Motion, I focused more to explain many of the Japanese cultural areas of uniqueness to enrich the understanding of where Aikido comes from. It is like trying to enjoy great Sake by knowing more about the type of rice grain, the quality of water of the regions, the brewing process, and tradition of the brewhouse, and the tremendous discipline and work effort of the Sake brewers.

The purely mechanical aspect of the explanation will have to wait for a visual format and I intend to work on it with some collaborators at a later date.

02 How to Define Aikido Technical Principles

In his book *Aikido Technique Methods* (合気道技法), written in 1962 and published by Kowado, Ueshiba Kisshomaru second Doshu describes the practical techniques of Aikido as follows.

The human mind and body are the products of nature created by the providence of heaven. Where these natural products blend into the movements of nature and become one, there is eternal progress and development of humankind. The power that is generated by learning Aikido is nothing but the power of nature hidden on one side of human beings, that by uniting a healthy mind and body with this nature, whenever it moves will always become a true technique, and will exert great power. If you compare the way O Sensei's Aikido is similar to a perfect sphere, the number of orbits in which it can move is innumerable, and if it differs by one centimeter, the movement of the technique can also become different. The technique embodies a technique that transforms into a different technique when the gap is 1 cm and again if the gap is 1 meter, the technique will turn again according to the 1 meter difference.

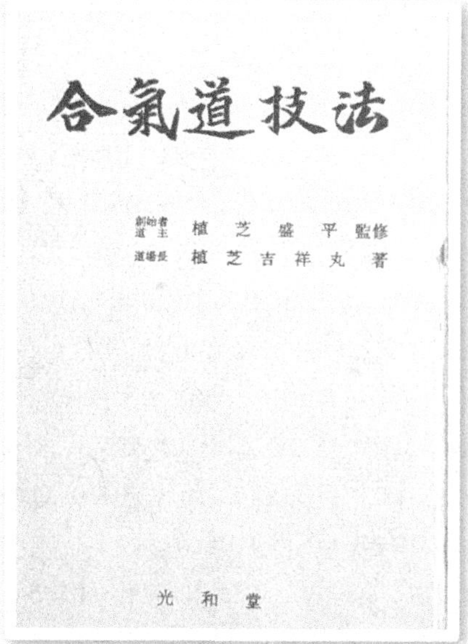

This is a very esoteric and somewhat poetic way to describe the art created by a genius like O Sensei. I can imagine that many of the contemporaries of Kisshomaru Doshu who witnessed the Aikido of O Sensei in person must have felt exactly like that. A similar description would apply to masters of many other arts, whether it is Michelangelo or Beethoven. When you reach a divine level, the layman will not have ways to describe with earthly words as the experience evolving in front of one's eyes is most likely beyond comprehension.

As mentioned earlier, I have lived more than two-thirds of my life outside Japan. While I fully comprehend the background description of Kisshomaru Doshu's statement, I also recognize that it would be extremely difficult to explain Aikido to an audience, regardless of where they come from, with this type of abstract description. My main occupational background has been in the auto industry. While you can use superlatives to describe a car, it is impossible to categorize the automobile in only abstract expressions. We need a yardstick to fully be able to compare and evaluate the characteristics of the car. I recognize that O Sensei, Michelangelo, and Beethoven cannot be described in static numbers, data, or even compared as art forms. However, for most of us to understand, we need more earthly terms to characterize the art.

To make matters more confusing, Aikido has other terms to describe key elements of the art such as **Kokyuryoku** (breathing power 呼吸力), **Kiryoku** (power of Ki 気力), **Hara** (belly 腹), or **Tanden** (body center 丹田), and so forth that are not quantifiable or describable. The translations in a foreign language may not relay its rather obscure (but important) meaning. Furthermore, as there are so many variables to the same core techniques, quite distinct variations could sometimes be defined in a rather loose category and have the same name. Cases in point are **Kokyu Nage** (breath throw 呼吸投げ) or **Koshinage** (hip throw 腰投げ).

The beauty of Aikido is that you train without weight class, age, or gender, and with no competition nor specific rules. When we train, we repeat the techniques over and over with our training partner. While the techniques may be the same at the implementation stage, the execution of the technique can vary depending on the size, weight, age, and other physical factors of the partners, or the speed, intensity, timing, and **Maai** (in-between distance 間合い) of the attack. You can train with people with disabilities as well.

The non-competitive element makes Aikido very adaptable to the partner and circumstances. As you adapt your training intensity to the level of your partner, Aikido has been recognized as an art that promotes empathy and collaboration. Naturally, when you wish to progress, it is always best to look for people with slightly more advanced levels to train with so that you experience more challenges for improvement. While it is understandable that most people prefer more advanced people to train with, in Aikido, we try to help beginners to understand the effectiveness of proper techniques.

Aikido is also defined as being an art of **Nagare** (flow 流れ) with less strict forms (**Kata** 型). Even within Aikikai, the breadth of interpretation between teachers is very broad. And even with the same instructor, the interpretation of the techniques may slightly—or dramatically— change during their different life stages. To learn the art when the standard is not defined very clearly, it would be best to stick to one teacher until you train yourself to understand the basic principles and establish your own criteria to evaluate the techniques.

O Sensei said that Aikido can create thousands of techniques, but only a few of them have been systematically compiled, and no book has yet been published or will be published that summarizes O Sensei's free and vigorous development of techniques.

Then the question arises as to how to define the technical principles of Aikido. Aikido is very deep and I profess I don't feel I know Aikido to the fullest. That said, with over 50 years of training, and many of those years under direct disciples of O Sensei, I have come up with some observations that may help some Aikido practitioners. My attempt here is to summarize what I personally learned over the years. As teaching criteria are somewhat broad in perspective, I feel there is a need to understand what the core requirements are to achieve good technical Aikido. In addition, there are many subtle elements that make the Aikido techniques flourish that many textbooks have forgotten to mention or overlooked. Also, I'd like to add many cultural or historical backgrounds to many of the Japanese martial peculiarities which are rarely explained or shared with non-Japanese Aikido practitioners.

In 1976, I received a scholarship to study in the US and applied to Harvard University Graduate School, where I was admitted. When I told Kisshomaru Doshu that I was going to Cambridge, Massachusetts, Doshu instructed Osawa Kisaburo Sensei (who was Dojocho at Hombu Dojo at that time) to write a letter of introduction to Kanai "Hambei" Mitsunari Sensei (1937-2004), who had moved to and was teaching in Massachusetts. I was a member of the Kisshomaru Doshu Souchou Geico (早朝稽古), Doshu's 6:30 morning class, so I was very fortunate to get this special treatment.

But it never occurred to me that this encounter with Kanai Sensei would have a major impact on my life as Kanai Sensei became my lifelong respected instructor of Aikido and much more. I had trained under numerous high-ranking instructors who were direct disciples of O Sensei, but it was through Kanai Sensei who had moved to the US to teach Aikido that I could say I was awakened to the deeper truth. Kanai Sensei at that time was a shy person with very few words, but his Aikido demonstrated phenomenal energy and powerful Aikido without brute force. The elegance and effectiveness of his moves truly impacted me. While he has had many capable students and many replicate his techniques, his effortless and powerful moves are beyond copying. He was another genius.

Kanai Sensei did not receive higher education in his life but to the amazement of all his students (many of us Japanese), he demonstrated a very high level of knowledge in history, metallurgy, philosophy, and art in general, but more than anything he surprised us by the fact that he had a copy of *Gray's Anatomy* in his study and his knowledge of human anatomy was beyond being an amateur.

He managed to give us some body/shiatsu realignment treatment if we were to get hurt somehow. I was deeply intrigued that Kanai Sensei had managed to be self-educated in such areas as anatomy. Later, I would find out why he was so deeply interested in this subject.

Kanai Sensei's inquisitiveness about human anatomy had a reason. When he came to the US back in the mid-1960s, he told us he weighed only 135 pounds (61 kg) and we know the majority of his original students in the Boston area were solid and heavier, easily weighing over 200 pounds. Being so small physically and only in his mid-twenties, he had no choice but to develop more effective ways to execute the techniques against his students who were much bigger, taller, and physically stronger than he was. So, his curiosity about anatomy became a major topic. Perhaps he would not have pursued anatomy if he had stayed in Japan.

He confessed to us that most of the **Uchideshi** (live-in students 内弟子) in Japan would fall asleep when O Sensei would talk about the divine philosophy of Aikido. The universal love and all the Shinto talk he shared was like praying to deaf ears, especially for the young Aikido practitioners. What impressed Kanai Sensei the most was that the petite O Sensei (who, as noted in *A Life in Aikido*, was only about 1.55 meters/5 feet tall but he weighed about 75 kg/165 lbs) could generate such a powerful Ki energy and throw people all over. This sense of energy being powerfully projected out became Kanai Sensei's eternal quest to find the mechanical secret that was behind this force. Kanai Sensei never showed too much interest in Ki power itself but always tried to find an explanation in how you could achieve that with mechanical force.

Once, when he visited Colorado before he passed away and we were driving to go to the hot springs in the San Luis Valley, he made a quiet comment about Tohei Koichi Sensei from Ki Society. He said that even though he highly respected his Aikido, he was happy that Tohei Sensei did not take over the leadership in Aikikai. He mused that if Tohei Sensei had done so, Aikikai would be promoting the concepts of the unbendable arm and focusing too much on Ki power. In comparison, the aspect of Aikido that mesmerized Kanai Sensei was that special feeling he got when he was being thrown by O Sensei. He especially was trying to decipher how O Sensei could muster that powerful energy from his small body.

The word that came from Kanai Sensei was the feeling of **Hajiki** (bouncing 弾き), which eventually becomes one of Kanai Sensei's interpretations of Hajiki Goshi techniques. While I accept that I cannot possibly cover all of the technical concepts that Kanai Sensei had cultivated, I will try to use some of his core principles and add some of my personal interpretations to look for a technical definition of Aikido. Although Kanai Sensei never mentioned it this way, my conclusion of the definition of the technical side of Aikido is that it is Balance in

Motion for the Nage while stealing the center of the opponent. I will continue describing in further detail as to why I believe this covers a lot of the technical principles of Aikido. Judo, Jujutsu, Sambo, wrestling and many others have similarities in the balance issue, but we will touch base on those elements that illustrate Aikido's unique characteristics that separate it from the other martial arts.

03 Toitsutai: The Integrated Body 統一体

In Aikido, we normally take a stance standing with one foot forward, either right foot or left foot. Most Aikido techniques start from this **Kamae** (preparedness stance 構え). This Kamae is executed in **Hanmi** (half body 半身). In this Hanmi stance, while the front foot is pointing straight forward, the foot in the rear might be placed with an angle outwards. When you are in Hanmi position as **Nage** (technique initiator 投げ), the tips of the hip bone positions turn diagonally facing the **Uke** (the attacker 受け). The same happens with the shoulders that are positioned diagonally vis-a-vis the opponent. It is a unique placement of the body that allows for **Tai Sabaki** (body movement 体捌き) (covered in the next chapter) with ease and at the same time allows Nage to be able to deflect attacks from the Uke.

When you look at Hanmi from a body structural point of view, the eyes and head are looking forward while everything else—the shoulders, hips, and feet—are positioned diagonally. This position is suitable for deflecting the opponent's moves, but is not ideal when you are about to counter the attack and you apply some throw or hold techniques. You could say that the Hanmi stance is a *transitional standing* stance. Hence, once you start the counter moves to an attack, you have to shift to **Toitsutai** (Integrated Body 統一体). We will delve deeper into Toitsutai now.

Toitsutai is not a term usually used by Aikikai people. I do not recall Kanai Sensei using the term in the 1970s but he started using it in the '90s as he was persuaded to write a book on Technical Aikido. The book never materialized but his thoughts were compiled into a video with the same title. While it is a highly recommended video, unfortunately it is not commercially available and does not cover all of what Kanai Sensei wanted to describe.

He passed away too soon before he could share more.

The word Toitsu (integration) has been used more by Tohei Koichi Sensei, who separated from Aikikai and established the Ki Society. To differentiate from Aikikai, Tohei Sensei focused on his philosophy of **Shinshin Toitsu** (body and mind integration 心身統一).

Kanai Sensei respected Tohei Sensei, but chose his focus to be more on the perfection of body mechanics to destabilize the opponent and not into some type of hypnotization believing the mind could control the body. If he was thinking about mindfulness, his perception of mind would have been a much more martial training, based to train the mind to react instinctively, anticipating the moves of the opponent. There was no time for the mind to be giving instructions to the body to react.

The concept Toitsutai that Kanai Sensei had in mind was the convergence of the power-producing physical elements moving in sync and focusing and projecting the energy output in one single direction. In other words, an Integrated Body or Unified Body. This would require that, at the final time of the execution of the technique, the toes, feet, hips, shoulders, arms, eyes, and head all would be looking in the same direction, and from there radiate the explosive energy of projection.

Toitsu, "integration," was meant to consolidate all the body elements. The mind is also present, but in Kanai Sensei's perspective, the mind was not a Zen-like concept. The desired result of Toitsutai only could be achieved by diligent and repeated training of the techniques where body and mind become one, not the mind controlling the body as Tohei Sensei taught. Perhaps Kanai Sensei went further. With sharp focus on achieving effective mechanical moves, pre-acting or reacting to the Uke's relative position moves, Kanai Sensei may have moved to the stage of Mushin (state of detachment of mind 無心).

Ki (or Qi in Chinese) is translated as energy or vital force. The Chinese character Qi 氣, from what I understand, comes from the hieroglyphics of "steam." As one can imagine from a steam engine, the steam energy can be maximized if the energy output is through one single outlet. If you had multiple outlets for the steam to go out, the steam would not be able to maximize its expanding force. Using a human body would be the same.

If you are executing a technique, you would want to converge all of your energy generated by body movements in sync and projecting out in one single direction. There are some exceptions to the rule, such as **Aiki Otoshi** (合気落とし) where you do not necessarily consolidate all the body parts heading in one direction as you execute the technique, but even in Aiki Otoshi, it would be desirable to reposition one's body in the direction where the opponent was thrown and projected to.

I was not trained in the medical field, but in order to highlight the magnitude of the impact of Toitsutai, allow me to play with some anatomy numbers. The human body is supposed to have, depending on how people count them, a range of 206 to 213 bones, 250 to 350 skeletal joints, 650 to 840 skeletal muscles, about 900 ligaments, and approximately 4,000 tendons.

Imagine if all these body elements worked in sync, relaying the energy and supporting other parts of the body to project the energy out. This would be the burst of power that Kanai Sensei felt when he was thrown by O Sensei. Having all the body elements working in sync would allow Aikido movements to be powerful but also very aesthetically beautiful.

Visualizing a cheetah running at full speed will give you a perspective of how the "wholesomeness" of the body is in action. Furthermore, cheetahs do not need to become mindful before going after their prey. I think whether you have your sense of Toitsutai or not, making use of your whole body makes a big difference in how much you may have mastered the art of Aikido.

Humans have lost many of the natural instincts we used to have. It becomes harder to depend on our senses, as many new environmental "noise" elements have popped up around us and we have become dependent on the inventions of modern society that replace many of our original instincts. For example, there are various man-made sensors designed to replace or augment human capabilities in specific tasks. These sensors often leverage advanced technologies to detect and measure physical properties, environmental conditions, or signals. Image, audio, proximity, motion, smell, and vibration sensors have greatly augmented or replaced human sensors.

Moving with awareness of the details of the body has become less needed for our daily lives, so it is by training in Aikido and other sports or martial arts that we try to reacquaint ourselves with our original instincts. You can still see our dogs and cats playing and rolling, falling, and getting up as nothing has happened to them. For humans, the same movement and activity can lead to catastrophic injuries.

But the basic functions are still there. Even if your head is not thinking about your body, your limbs are moving firmly in tandem, enabling you to perform your very basic target movements. When one body part moves, the rest of the body moves in a chain reaction. In fact, the body is mostly made to move automatically. It may be easier to understand if we say that the body is interlocked. Expressed from a joint perspective, this is called a "kinetic chain."

In Aikido, being the Nage requires one to move freely to redirect the energy of the Uke, and that means one rarely stays static in one position. As we do not move our muscles in isolation and need to stay in balance and with good posture, a kinetic chain takes place. A typical move in Aikido can include foot placement and grounding to knee and leg activation, proceeding to hip rotation entering into core engagement, and applying shoulder and arm alignment. Traditional bodybuilding is usually designed for muscle isolation, but the beauty of Aikido is that one needs more overall functional movement designed to work a lot of muscles together.

When one part of the body moves, other parts of the body move under the influence of that movement, which in turn move in tandem, causing other body parts to move, and so on, creating movement in a chain-like manner. The shape of the bones and the structure of the joints propagate movement in this way.

We can look at the muscle chain that occurs in response to gravity as well. Let's imagine the body relaxed and seated in a chair. Your pelvis has fallen back and your hips and back have bent. Now, try pulling up the top of your head toward the ceiling. How do you feel? Hasn't your pelvis risen on its own and a light arch (kyphosis) formed in your lower back? This is the chain of anti-gravity. When the body tries to stretch to overcome gravity, even the pelvic muscles are switched on and move from the tip of the head to the hips. This anti-gravity kinetic chain is the most important thing that makes the body connected.

We have talked about how the body generally moves in an automatic way. This occurs especially in the case of basic movements such as walking. However, there are postural requirements for this kinetic chain. These are the following three points:

1. The head is extended toward the ceiling.
2. The pelvis is in an upright position.
3. The body is not tilted back and forth or from side to side (i.e., it is in the center of the body).

When these postural elements are in place, the body responds well and the proper kinetic chain occurs, even if you perform the intended movement without thinking about the body. Conversely, if these are not in place, the proper kinetic chain will not occur.

In other words, in a posture where the head is drooping, the pelvis is collapsed, and the body is tilted back and forth or from side to side, the proper kinetic chain does not occur, and the body is subjected to unbalanced stress. This is shown in the drawing on the left.

If this usage continues, joints and muscles become tired and broken. This is how many lifestyle-related osteoarticular (orthopedic) diseases are formed.

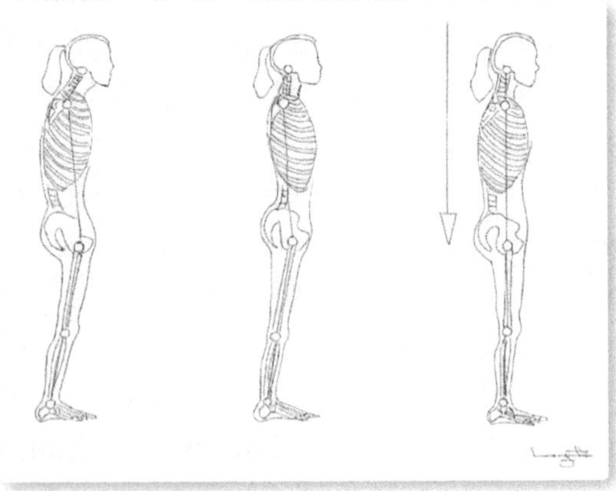

Many people go to a doctor after their knees, hips, or shoulders start hurting, saying, "I'm in pain. My body hurts."

However, often they are told that it is aging, and they are given compresses and medicines. The real cause, however, is the breakdown of the kinetic chain. In other words, it is posture.

Movement actually comes from correct posture, and this is what is important. Proper kinetic chain is a condition of correct posture. Therefore, posture and movement are not separate concepts, but are essentially one and the same. The same is true of the mind and body, which we humans divide as separate concepts, but in reality they are just different perspectives of the same thing. What makes Aikido a phenomenal exercise without any type of equipment is that most of the time we have to interact with a partner by destabilizing the partner while maintaining our own balance, all in very active motion.

This brings us back to Toitsutai. Aikido repetitive training allows us to find elements beyond proper posture. Toitsutai has some additional benefits. As you can see, bringing most of the joints, muscles, ligaments, and tendons in sync improves the flow of energy many times more. Not only that, it allows us to project out more energy without stressing some parts of the body. The various Aikido moves done in Toitsutai form will also maintain great balance. Here lies one of the secrets of Balance in Motion. As all the body elements are in sync, the Nage has more room for flexibility as well.

Going further, after you as Nage have executed the technique and you have achieved the throw or pin down, Aikido's technical completion is achieved only when you have the proper level of **Zanshin** (unrelaxed alertness 残心). The characters of Zan Shin mean "remaining" "mind/attention." It would be almost impossible to conduct perfect Zanshin without completing the technique in Toitsutai form. If the total body composition in its final stage is not converged or focused, how can you exert Zanshin? In later chapters, we will cover some of the key training elements to train yourself to always approach matters in a Toitsutai form.

04 Tai Sabaki (Irimi, Tenkan, Tenshin, Nenten)

When one starts training in Aikido, the first moves most likely taught will be the step movements we take such as **Irimi** (entering, sliding in 入身), **Tenkan** (side backward turn 転換), and **Tenshin** (stepping back 転身). So we could say that the shift in one's position vis-a-vis the opponent forms the introduction to Aikido and it is very likely that most dojos may practice these moves with or without a partner later on as part of warming up, even for advanced practitioners.

These defensive body moves preceding the counter techniques are called **Tai Sabaki** (body movement 体捌き) in Japanese. In other Japanese martial arts, it is also called **Ashi Sabaki** (feet movement 足捌き).

The word **Sabaki** has a special nuance. Here as a matter of convenience we are translating it as "movement," but it has a much deeper meaning to it. Sabaki has a feel of expert handling, processing, disposing, wielding, and many more. As an example, while anyone can cut a fish, a professional sushi chef would "sabaku" his fish in a masterly way, slicing and preparing the fish slices for sushi. So when you use the Sabaki of the body in Tai Sabaki, it has an implication that you are moving your body "masterly" to achieve the desired results in your Aikido moves. At least, that is the meaning of the Kanji character.

When we ask why these moves are so critical for Aikido training, it is because these moves allow the Nage to reposition one's body vis-a-vis the opponent Uke so that one can avoid being in a receptive end of an attack angle or at least shift to a position which will allow the Nage to be ready to use the Uke's energy in a technique. The technique may involve a joint technique, a throw technique, an immobilization technique, or something else while destabilizing the Uke.

Tai Sabaki often is accompanied with **Kuzushi** (destabilization 崩し) movements. **Atemi** (a blow/strike to prevent Uke from regaining attack stance 当身) can be part of the Tai Sabaki as well.

As important as these body shift moves are, the repetitious movements become something that is taken somewhat for granted and not many people attempt to explore the ultimate form of Irimi, Tenkan, and Tenshin moves. While these three shifts (Irimi, Tenkan, and Tenshin) are very important, we also need to take into account that the moves should be done in a way that the executioner of the move (Nage) can maintain utmost stability during rapid movement, shift the body with firmness, and come to a position to cause destabilization (Kuzushi) of the opponent (Uke) with strong martial stances.

Another body movement that does not require shifting of the feet position but is extremely crucial is the **Nenten** (torsion or gyration 捻転). This Nenten is executed by rotating the feet/body up to 180 degrees (in the opposite direction) without taking extra steps. I have seen that some people call this **Kaiten** (turning 回転) or **Jiten** (pivot 自転) or **Tenkai** (turning 転回) as well. This body movement is used often to merge with the energy of the opponent.

All the step movements—Irimi, Tenkan, and Tenshin, and even the gyration movement Nenten—require firm grounding as well as agility. So, we could say that before learning the basic moves, we need to train the muscles of the feet that allow us to maintain strong balance while taking actions.

Not many people know that one quarter of all the bones in the human body are in our feet. As you learned in Chapter 3, the human body has about 206 to 213 bones. Each foot has 26 bones, 33 joints, and a complex network of more than 100 tendons, muscles, and ligaments that keep our stability and balance. So, it's no wonder that when these bones are out of alignment, so is the rest of the body.

In Kendo, there is a saying that highlights the priority of important points in training. They call it **Ichigan Nisoku Santan**

Shiriki (一眼二足三胆四力). They rank by importance in this order:

- First: eyes (observation, insight 一眼)
- Second: foot movements (Ashi Sabaki 二足)
- Third: courage, bravery, and valor (Tan 三胆)
- Fourth: muscles (Riki 四力)

As the ranking goes, the foot movement is more important than valor and muscles in Kendo. Kendo goes into different foot movements such as **Suri-ashi**, **Okuri-ashi**, **Tsugi-ashi**, **Ayumi-ashi**, **Hiraki-ashi**. Interesting to note is that Kendo has so many different foot movements but nothing like the Tenkan movement of Aikido. People may perceive that training with the Shinai (bamboo sword) would be more important than foot movement in Kendo. As this saying shows, however, even in Kendo the movements of the feet are much more important. One could say that it would be the same in Aikido with the importance of the foot movements.

Tai Sabaki moves have to be executed with the intent of transitioning into Toitsutai, which was explained in Chapter 3. This is especially true if one does a combination of Tai Sabaki moves. For example, you may do Irimi, Tenkan, and finish with another Irimi, as you should end with a clear Toitsutai and project energy out. We will discuss more in the breathing chapter, but the final Toitsutai move could coincide with breathing out.

The Tai Sabaki moves are important in that you make the movement first to avoid being the target of an attack, second to move to a position where you are safer, and third to allow you to start your moves to destabilize the opponent. The foot moves are critically important as they make the upper body Sabaki much more stable and easier so that the Nage does not lose balance. Here again is another small hint of the Balance in Motion. Where you move, how you deflect the attack, and how you destabilize the opponent are a package deal.

I will not go over the Tai Sabaki moves in too much detail, as countless books and videos cover it. However, it is important to highlight that when you do the solo practice or with a partner, you keep those points above in mind and also try variables by combining with height adjustments by lowering your center in good posture using your knees—not by bending your back. It doesn't have to be the same height all the time. Try to generate waves in height to further destabilize your opponent. I also like to train tilting my body axis, imagining throwing the opponent and creating some counter weight to achieve destabilization of the opponent.

Now the last, but very important, element: Tai Sabaki has to be considered as one crucial technique of Aikido martialness. As we do not have competition, there are times that the martialness aspect of Tai Sabaki is forgotten. Many Japanese Aikido books and teachers, including Kanai Sensei, raise the issue that Aikido is a martial art based on relative interaction. You need an opponent or partner to train. In some ways, it is a small universe of interactions. Each training opponent will react differently, but the training has to be based on intensity for both sides to progress.

For training, it is best to change the criteria of the attacks by changing intensity, strength, and speed with credible attacks so that the Nage can also learn to deflect using the Tai Sabaki. Of course, this spirit has to be taught in class but the instructor will have to take into account the mix of the levels of proficiency, as there is no need to attack with too much intensity to someone who is a beginner or has some difficulties.

05 Shisei (Posture) Impacting the Center of Gravity

Whether it is tea ceremony, flower arrangement, or the theaters of Noh and Kabuki, one common thread that these Japanese cultural arts have with **Budo** (the Martial Way 武道) is good **Shisei** (posture 姿勢).

This is especially true, as all of these arts require people to sit on the floor on the tatami and have to stand up effortlessly and elegantly without a hint of losing balance. Unlike in Western culture, the Japanese definition of good posture starts with a harsher requirement, as good posture begins from **Seiza** (squatting straight [also correct sitting] 正座). A proper Seiza posture trains you to keep the **Senkotsu** (sacrum bone 仙骨) in a vertical mode, which improves your overall upper body balance and posture.

Similarly, for Aikido people who train in **Shikkoh** (knee walking 膝行), the proper way to knee walk is to maintain a very vertical sacrum bone, achieving good upper balance with overall stability.

Those that can conduct the Shikkoh move effortlessly in a balanced mode are likely to be able to conduct the same while standing. As the base in Shikkoh walk becomes a well-balanced tri-pointed leg exercise (two knees and one tight-together feet), this becomes the ideal lower body platform to place the upper body in a relaxed mode. A good Japanese Shisei—both standing and from Shikkoh—is relaxed posture, with relaxed shoulders but with a firm and elegant presence. If you stand while focusing on keeping the sacrum bone in a very vertical position, the upper body falls into place ready for action.

A good Shisei can be achieved without the sharp edges that are seen in Western military tradition. With the military "standing at attention" posture comes the image of a posture in which the buttocks are extended and the chest and shoulders are stretched. It is precisely the type of man-made beauty.

The military created a standard of posture in which elements of the body are structured in a neat, precise, and uniform way; you could even say in a geometric pattern. Perhaps you could also say that it is like a contrast of a Western-style garden next to a Japanese garden, which is a reproduction of nature in a miniature form, and not based on man-made patterns. Western-style gardens stress the beauty of artificial creation and the skills of mankind.

It's hard to keep a tense posture for a long time and it is not a practical posture for someone who may have to take defensive action in an instance.

Let's take a closer look at the "standing at attention" posture. The position "at attention" is a military posture which involves the following general positioning:

Standing upright with an assertive and correct posture: famously "chin up, chest out, shoulders back, stomach in."

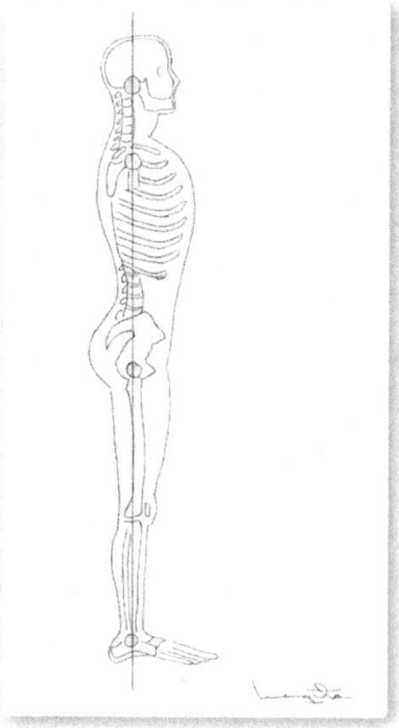

- Arms fixed at the side, thumb or middle finger parallel to trouser or skirt seam, depending on military drill specifics.
- Eyes front: head and eyes locked in a fixed forward posture. Ideally, eyes unmoving and fixated on a distant object. Blank facial expression.
- Keep the heels together, with the toes apart and the feet at a 45-degree angle.
- No speech, facial, or bodily movements except when as required by military drill.

— Source: Wikipedia

On the other hand, the posture emphasized by martial arts is achieved by trying to raise the pelvis. As described earlier, the "sacrum" is brought to a vertical angle. This posture becomes more like the "cheetah" on guard with no stiff muscles and ready to attack. By doing this, one will be conscious of having an axis in one's body. Even if you keep this posture for an extended time, you will feel a sense of stability and not feel tired. Also it allows you to do any Tai Sabaki at a spur of a moment.

When a person is in high tension, the upper body gets very stiff. When you see someone with stiffness and high tension, you may say "relax your shoulders and take a deep breath!" When you tighten your muscles in order to take action, you inadvertently make things harder, which leads to the inability to exert force in an emergency. Muscle tension is the state of a muscle that is contracted. This can be caused by a number of factors, including stress, physical activity, or repetitive motion. Muscle tension can lead to a number of problems, including pain, stiffness, and reduced mobility. In martial arts, you don't wish to reduce your mobility, so you should drop your shoulders, relax your chest, and consciously stretch your back.

People who have been practicing martial arts for a long time are said to have a good posture, but it is not the rigid posture of the guards in Buckingham Palace. The good posture of martial arts practitioners is not just a part of etiquette, but it can be said that the martial artists who have challenged the game for life and death have reached the point of Shisei. The results of seeking a highly functional posture and resulting attitude without *waste* are brought together in that decisive moment.

Let's continue with the benefits of good Shisei. The word Shisei is a combination of two Chinese characters: **Sugata** (appearance/shape 姿) and **Ikioi** (momentum, vigor, force 勢い). As you might imagine, for the ruling Samurai class, good **Shisei** was essential and an attribute coming from a Samurai class with proper **Shitsuke** (discipline 躾). Incidentally, the character for discipline is a combination of two letters that mean body (身) and beautiful (美).

Beginners in martial arts, for example, often process things in their heads before moving. When they try to perform a technique, because of the extra processing, their movements become awkward and often their postures are compromised. It is desirable to have a stable posture form, in which the core is solid and the rest of the body moves freely, like keeping the chakras open but firm. For this reason, a good posture that leads to flowing movements will lead to faster progress in martial arts. Interestingly, the sense of all-out effort and hard work disappears when rational body movements based on good posture are acquired. The technique moves look simpler, elegant, and much more fluid. Even more strange, when one has good posture and has fully mastered the movements of Budo, the movements appear to be flowing, and one gets the illusion that the technique is even moving more slowly. A great flow of Ki (**Ki no Nagare** 気の流れ) is achieved when your body reacts automatically against the move of the Uke without processing in your head, and it starts when you establish a strong body core with impeccable posture.

As in daily life or in martial arts training, bad posture may often lead to physical injuries and pains. Bad posture leads to constricted blood circulation where there is extra muscle tension. We all know what can happen if you spend long hours in front of computer keyboards looking down. On the other hand, proper posture has many additional benefits for your martial arts practice. The techniques start to look more effortless and less muscle tension conveys less aggression to the opponent, which causes less resistance. Your reaction to the opponent's moves becomes faster, and the probability for injuries is much diminished. Of course, you also waste less energy and can enjoy training even more.

Perhaps we can learn from the master swordsman Miyamoto Musashi, who compiled a list of things that are important to have in attaining correct posture for fights. His conclusion was even more impactful, as it has become a major mantra of the importance of good Shisei:

Stance (read Shisei) in Strategy: Adopt a stance with the head erect, neither hanging down, nor looking up, nor twisted. Your forehead and the space between your eyes should not be wrinkled. Do not roll your eyes nor allow them to blink, but slightly narrow them. With your features composed, keep the line of your nose straight with a feeling of slightly flaring your nostrils. Hold the line of the rear of the neck straight: instill vigor into your hairline, and in the same way from the shoulders down through your entire body. Lower both shoulders and, without the buttocks jutting out, put strength into your legs from the knees to the tips of your toes. Brace your abdomen so that you do not bend at the hips. Wedge your companion sword in your belt against your abdomen, so that your belt is not slack — this is called 'wedging in'.

— *The Water Scroll. The Book of Five Rings*, by Miyamoto Musashi, translated by Thomas Cleary (Barnes and Noble), 1993.

Musashi takes the issue of Shisei (posture) so seriously that he leaves a quote that shows how much he valued it:
In all forms of strategy, it is necessary to maintain the combat stance (Shisei) in everyday life and to make your everyday stance your combat stance (Shisei).

Most people who reach the level of Shodan and start wearing **Hakama** (pleated pants 袴) may experience a new feeling of support in the lower back. Most Hakamas have what is called a **Koshiita** (lower back board 腰板). The Koshiita is typically made of a lightweight material, such as wood or plastic, and is covered with fabric that matches the rest of the Hakama. The Koshiita helps to keep the Hakama in place and prevents it from riding up, and it also provides support and protection to the wearer's lower back. When you take some long vacations from training and come back to Aikido, the feeling of having the Koshiita support is a nice reminder of keeping the sacrum upright. In this way, the Hakama is something that looks like a cosmetic ornament but it comes with a wonderful reminder system of your posture.

Good posture is one of the major prerequisites to achieve Balance in Motion in Aikido. Just simply take a look at the masters. For us Aikido practitioners, we can see the videos of O Sensei, Kisshomaru Doshu, current Moriteru Doshu and Mitsuteru Dojocho, and we cannot admire enough the beautiful Shisei they use for demonstrations. We need to realize that good posture creates a powerful and stable axis that can rotate efficiently, maintaining the Balance in Motion allowing the Nage to perform the techniques effectively and efficiently. You can see pictures and videos of people proudly performing techniques with their backs bent, but from the point of view of rational Aikido mechanics, such Aikido forms are not ideal. We should remember the comments by Miyamoto Musashi and/or emulate the beautiful moves of Doshu and Dojocho.

奇山秋
乙酉暮月

06 Big Role of Big Toe and Toes

When we look at human evolution, we start to understand the importance of the foot structure and the role that the big toe plays. The primates have feet that are suitable for climbing trees and many anthropological studies have uncovered the remains of early Homo sapiens showing changes in the bone structure of the feet that enabled them to be able to walk on two feet, without the support of the arms. And the major contributor to this new ability was the change in the front of the feet, the big toe in particular (metatarsals and phalanges). As they say, walking upright is one of the defining features of human lineage and it required major changes from the great ape foot structures.

Perhaps modern humans feel that it has always been like this and we take for granted that we can walk upright. We have to recognize the phenomenal transformation that took place to achieve it. The change to upright walking required a huge wiring and support structure to be able to keep the balance. To walk, run, or jump, the front of the feet was transformed to cope with new necessities.

Fascia, tendons, and ligaments are ionically charged and receive sensory information providing feedback to the brain, which enables us to move with skill. To highlight the complexity of the foot, here are just a few of the structures involved:

- Over 7,000 nerve endings in the sole of the foot (no wonder feet are among the most ticklish parts of the body or targets of painful and vicious tortures)
- 26 bones in each foot (14 bones on the front of the foot alone) that work with intrinsic muscles to provide a static or dynamic balance
- 250,000 sweat glands that can produce 4-6 oz of perspiration a day

in active feet (This is thought to have helped our ancestors to increase the grip of feet on the ground when running from predators or any other threat.)

- 33 joints, and over 100 muscles, ligaments, and tendons.

Several years ago, I came across an interesting Russian study after searching on the internet. The study was conducted by Novosibirsk State National Research University and the Scientific Research Institute of Physiology and Basic Medicine. It was reported as "Martial arts can help senior citizens improve their balance."

According to a *Russia Beyond* article, "Researchers asked 35 women in menopause (ages 54 to 78) to get up and then sit down again with open and closed eyes. The best results were obtained by women who during the last eight years practiced aikido a few times a week. Aikido is an exercise that puts pressure on the sole, especially on the big toe."

It is interesting to note that the study compared people who had similar training in fitness, yoga, and dance, and it showed that Aikido brought the best result. (*https://europepmc.org/article/med/28476511*)

Naturally the feet and toes are part of a complex system connected with all parts of the body and the brain. As mentioned previously, the feet have over 7,000 nerve endings in each foot. All these nerve endings have a purpose, and the big toe is connected to the inner thigh muscles and subsequently to the brain where the balance is controlled. Aikido uses the big toe, with all the Tai-Sabaki (体捌き) movements and Shikkoh (膝行) for proper posture, so it's no surprise that Aikido practitioners have better balance than those who do other forms of movement. Think about this. In English, when you say someone or something "**keeps you on your toes,**" you mean that they cause you to remain alert and ready for anything that might happen, even when you are wearing tight encasing and hard sole shoes.

A good practice for training the toes and base of the big toe in Aikido is **Zagi** or **Suwari-waza** (sitting technique 座技 or 座り技). O Sensei apparently was very satisfied if people were training in Suwari-waza.

Since the sitting technique moves with the toes upright, the toes and feet front bases are trained. Also, in pinning techniques from Suwari-waza, the opponent is finally pinned face down on the tatami mat with the Nage (technique initiator 投げ) executing the pinning technique sitting on his/her heels with toes upright. Therefore, the proper form of concluding the technique is the Nage sitting with good posture applying the pinning with his or her toes upright. Concluding the technique sitting flat does not provide benefits to toe training.

Strengthening the big toes, therefore, allows you to improve your overall balance, and this is essential for the execution of a balanced technique—whether from standing or sitting. As we have seen in the chapter about Tai Sabaki, imagine yourself moving with a vertical axis that goes through your body from the top of your head to your toes with the ground pivotal point basically being your big toe or the front end of your foot. Instead of gyrating using the heel of your foot, if you use the tip of your foot, you can definitely extend the radius of your move and thus achieve a much stronger centrifugal force. If the axis is wobbly, no technique can be effective. The basis of Aikido is all about maintaining this centerline.

Benefits of Toe Training

To begin with, toe strength is said to decline steadily with age. This means that as we age, the need to strengthen our toes increases. So, how does toe strength affect physical health?

(1) Prevention of Falls

In fact, toes play a very important role in maintaining the ability to balance. When a person is about to fall forward from a standing position, the toes flex (bend) to grasp the ground and maintain the body's balance. Conversely, when the body is about to fall backward, the toes bend upward (extend) to maintain the center of gravity so that the body does not fall. Therefore, training the toes helps prevent falls.

(2) Improvement of Walking Speed and Physical Fitness

When a person walks, he or she grips the ground with the toes and kicks out behind to generate propulsive force. When the toes are strong, walking is more stable and the feet do not tire easily even when walking long distances. In addition, when the toes are strengthened, we are less likely to fall over and can walk long distances with comparative ease, which in turn helps to maintain physical fitness.

(3) Improvement of Flat Feet

Flat feet are a condition in which the soles of the feet have little lift in the arch area. Flat feet prevent the natural cushioning function of the feet from working properly and cause fatigue, as well as many other complications, including Achilles tendinitis, foot pronation, ankle swelling and pain, and shin splints. Strengthening the toes strengthens the natural function of the feet and helps to improve flat feet.

Let's not forget about the toenails. Toenails play an important role in walking. Immediately before lifting the foot, the toe always kicks the ground. At this time, the toes also support the body weight. The toes receive force from the ground through action and reaction. Without toenails, the toes would not be able to kick the ground and would not be able to receive enough force from the ground.

If the nails are cut too short, the force received will cause the nail tips to dig into the skin. Ingrown nails may develop for many reasons. Some cases are congenital—the nail is just too large for the toe. Trauma, such as stubbing the toe or having the toe stepped on, may also cause an ingrown nail. However, the most common cause is **tight shoe wear** or improper grooming and trimming of the nail. (Source: American Academy of Orthopaedic Surgeons)

Japanese Footwear: Form to Function

Because the shape and design of footwear is so important, now allow me to divert the topic to Japanese traditional footwear, as it is closely connected with the big toe and feet.

Traditionally, most Japanese footwear had a split toe design, with the exception of straw snow boots and some ceremonial shoes for the nobility. You could say that the split toe design shoe was ingrained in the Japanese lifestyle, especially for the ruling Samurai class. Another characteristic is that Japanese footwear never developed the concept of heels.

One of the first Westerners who chronicled Japanese lifestyle and shoes was a Jesuit named Luis Frois, who observed this in 1585: *"Our shoes, boots and slippers have soles that are separate (from the uppers): Japanese tabi have no separate sole and are made of one*

continuous piece of leather." Historians debate if the tabi shoes were made of leather, but the most important observation follows: *"In Europe it would be ridiculous to wear shoes with half the foot sticking out: in Japan this is considered most stylish and only bonze (monks), women and the elderly wear shoes that fit the entire foot."*
— Source: Quotes from *The First European Description of Japan, 1585: A Critical English-Language Edition of Striking Contrasts in the Customs of Europe and Japan*, by Luis Frois, S.J. (Japan Anthropology Workshop Series), and part of "Japan Anthropology Workshop" (31 Books), by Luis Frois SJ, Daniel T. Reff, et al. Mar 14, 2014.

The footwear Luis Frois observed in the "half the foot sticking out" shoe was most probably a footwear style called **Ashinaka** (half-foot sandal 足半), as can be seen from this historic illustration.

This Ashinaka sandal can also be observed from the folding screen illustration of the Mongol invasion in 1281, so it is something that Samurais must have used for a long time.

The Illustrated Miracles of the Kasuga Deity (Kasuga gongen reigenki emaki), in 20 fascicles, is an illustrated hand scroll that contains depictions of the miracles of the Kasuga Deity, the tutelary god of the Fujiwara clan. The deity was worshiped at the Grand Shrine of Kasuga (Kasuga Taisha 春日大社), which wielded great political and cultural influence along with Koufukuji, with which the shrine was integrated. The scrolls were completed in 1309 (Enkyou 2), during the late-Kamakura period, by the painter Takashina Takakane.

Oda Nobunaga, a leading war lord (1534-1582) during the Warring States period, usually carried an extra Ashinaka sandal pair on his waist obi belt. In the 1573 Tonezaki battle with the Echizen war lord Asakura clan, one of his lieutenants lost his zori sandal while struggling with an enemy. Even so, he was able to succeed in taking the head of the other military commander and brought it to Nobunaga. Nobunaga saw his lieutenant Kanematsu Masayoshi (兼松正吉) with injuries but still managed to bring the head of the enemy even while he had lost his sandals. As a reward, Nobunaga gave Kanematsu the Ashinaka that he was carrying. This Ashinaka was kept as a family treasure and is now preserved as a collection piece at a museum in Nagoya City. (*http://www.museum.city.nagoya.jp/hideyoshi/index.html*)

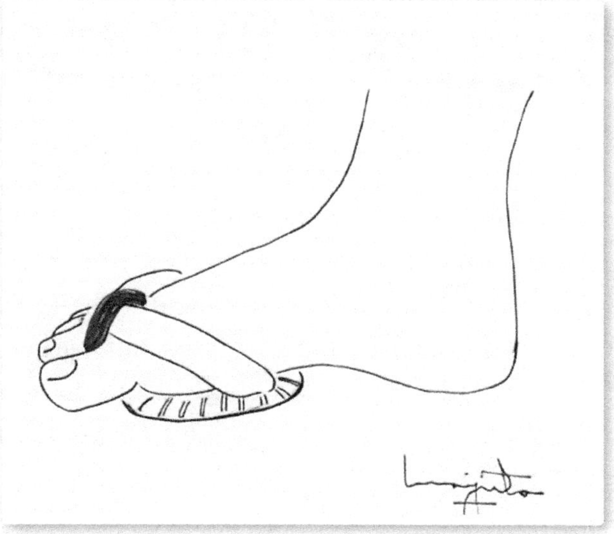

Interestingly, not many people know this but Ashinaka is even sold in Japan today. You can search at amazon.co.jp (search word: NHK Men's Half Zori Sandals With Nose Straps). Promoters of the Ashinaka tout the benefits as follows:

Simply by wearing the Ashinakas, unnecessary tension is easily released, and the shape of the Ashinaka helps to develop the three arches of the sole. As a result, foot problems can be prevented and improved. In addition, the weight is naturally shifted backward, the posture becomes closer to the neutral position, and the muscles that are supposed to be working can move more easily.

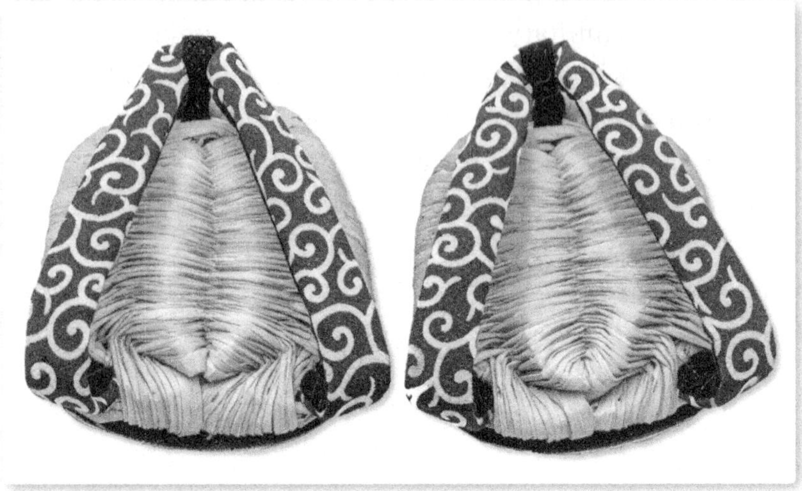

The health and beauty industry in Japan continues to promote this trend and sell products based on the traditional concept of Ashinaka. For example, they promote the benefits as diet slippers called Bikyaku (beautiful legs 美脚), indoor acupressure healthy sandals. Even in the era of Westernized shoes, some Japanese traditions are alive.

Another interesting observation comes from an American writer named Lafcadio Hearn, who wrote a book in 1894 called *Glimpses of Unfamiliar Japan* (available free online). Hearn, who later became a naturalized Japanese citizen and wrote many novels, gave his first impressions on Japanese footwear as follows:

The tabi, the white digitated stocking, gives to a small light foot a mythological aspect—the white cleft grace of the foot of a fauness. Clad or bare, the Japanese foot has the antique symmetry: it has not yet been distorted by the infamous foot-gear which has deformed the feet of Occidentals.

But one observation that highlights the Japanese way of walking is that Japanese used the front of the foot to walk. He writes:

The whole population is moving on tiptoe, and the multitudinous twinkling of feet over the verge of the sunlit roadway is an astonishment. All those feet are small, symmetrical—light as the feet of figures painted on Greek vases—and the step is always taken toes first; indeed, with geta it could be taken no other way, for the heel touches neither the geta nor the ground, and the foot is tilted forward by the wedge-shaped wooden sole. Merely to stand upon a pair of geta is difficult for one unaccustomed to their use, yet you see Japanese children running at full speed in geta with soles at least three inches high, held to the foot only by a forestrap fastened between the great toe and the other toes, and they never trip and the geta never falls off.

If you were to wear **Geta** (wooden clog with a strap between your big toe and the rest 下駄) and land from your heels, the Geta is likely to slip away and perhaps the same for Ashinaka sandals. So for more than a thousand years, the Japanese must have used the front of their feet to walk. As historical master swordsman Miyamoto Musashi said in his recommendation, the proper Shisei was: "*Lower both shoulders and, without the buttocks jutting out, put strength into your legs from the knees to the* ***tips of your toes.***" Conceptually, the military and sports in the West had similar understanding that being **on the back foot** meant *to be put in a defensive position, to be in retreat, to be knocked off balance.*

I am going to digress a bit, but recently I had the opportunity to view a historical exhibition at my alma mater Keio University and one photo caught my attention. Keio University, which was very active in incorporating new Western trends into Japan, had a display photo of some ice skating shoes used by Keio sportsmen in the late 19th century. The ice skating shoes were made from Getas with some blades added to skate.

As shown in the photo, when we look at traditional Japanese **Waraji** (straw sandals 草鞋), the strap is at the edge of the sandals exposing the toes to the ground. This surely strengthens the tip of the feet and allows the Japanese to not only have better balance but also

run faster and be more agile during combat. I also believe that the exposure of the toes to the ground fortified the skin so that the toes were not as delicate as our modern toes and feet soles are.

We are always amazed by children in less developed countries who can play barefoot on rocky grounds but we forget that we have spoiled our feet. As Japanese wore Waraji sandals until the mid 1800s, it is only natural to assume that their toes were exposed to most surface conditions. For the Samurai, the mobility of the big toe was crucial so part of the foot armor worn by Samurais had unique feet protection named **Kohgake** (foot instep guard 甲掛).

We have to thank these non-Japanese for giving descriptions of the footwear and the observation of the way Japanese walked centuries ago, since Japanese did not record the way they walked. As Japan westernized from the late 1800s, many of the traditional customs have become extinct among the general public.

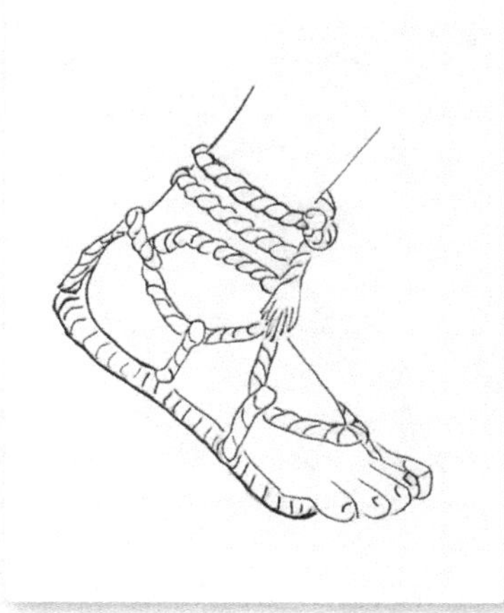

— Source: photo of the Kohgake restoration used with permission by David Thatcher

Perhaps some of the early practitioners of martial arts kept that tradition as well as the performers of Noh and Kabuki theater that follow long traditions.

As Aikido is based on traditional Japanese arts, it would be safe to assume that many of the Tai Sabaki moves need to move forward, pivot, or stop using the front of the feet. If you do Irimi, it would be obvious that you will step forward from the tip of your feet. Or as you pivot to the side in Tenkan, you will place some weight onto your front toes, mainly the big toe. If you step back in Tenshin, you will stop by extending your leg and using your front toes to make a stop. While we will talk more in the next chapter, the most energy efficient way to move forward is to tilt your body axis forward and let the foot/leg go forward. You will see that you will be stepping forward from the tip of your toe.

07 Aikido Steps and Namba Steps

When one starts Aikido, one of the first challenges is how to get used to the steps of Aikido. It is a logical concept as it is but if you try to do it flowingly, it may get very confusing. It is confusing since in Aikido we move each body side as a unit and is very different from our normal body moves. Let me explain it further to add clarity.

In Aikido, the Kamae (preparedness stance 構え) that we take synchronizes the lower body with the same side upper body. For example, right arm with the right leg, or the left arm with a left leg resulting with minimum torsion of the waist. When we train the basic Tai Sabaki (body movement 体捌き) moves, most of the time even the beginners do it well. But as soon as we start some techniques, most people suddenly forget to apply the side synchronization. The upper body and the lower body go out of sync. This indicates that Aikido moves are very different from our day-to-day body moves. But this is not unique to Aikido. Aikido's moves were strongly influenced by the sword techniques and as such have some similarities with many other martial arts, especially those that originated in Japan.

When comparing Japan and the West, it's not only walking but also mannerisms and other non-verbal expressions, as well as the values of standing and sitting, that are fundamentally different. That is why it drew the attention of the Jesuit priests that landed in Japan in the late 1500s. These differences are not regarded as mere differences in customs, but as differences in "reason" in the pursuit of rationality and ideals, where the values pursued by each are different. The differences originate in the social framework, way of life, clothing (including footwear, as mentioned earlier), and many other elements. So it is not only what is different but why it is so different that becomes a point of consideration. If this impacts the moves of Aikido in some ways, we need to understand the total social background as well.

Based on Samurai tradition, a school of etiquette emerged in Japan in the late 1100s and that has continued for close to 1,000 years to dictate to some degree what is acceptable in social interactions with people. The **Ogasawara school** (小笠原流) laid the foundations for etiquette for the Samurai class of Japan. These rules and practices covered bowing (the school's teachings describe nine different ways of performing a bow), eating, marriage and other aspects of everyday life, down to the minutiae of correctly opening or closing a door.

Based on the Ogasawara-style teachings, walking is a complete **Suriashi** (sliding step すり足), moving legs forward with both heels and toes of the feet never leaving the ground. As part of the formality, the Japanese-style standing posture puts both hands on the thighs, bends the knees lightly, lowers the hips slightly, and keeps the upper body vertical. The legs should face forward rather than outward, and the distance between the legs should be a natural waist width.

To walk from this position, slide one leg forward with your leg, bend the knee of that leg, and move your upper body over your knee. This movement becomes the driving force (stepping drive). Then, with the heel of the remaining foot on the ground, it can catch up with the foot that was put out in front. While walking, keep your upper body fixed and keep your hands on your thighs (do not swing your arms). Knees are not always fully stretched, but the hips are slightly lowered to eliminate vertical and horizontal movements. Also, make the proper timing of breathing to correspond to walking.

In the case of outdoor walking, instead of a complete Suriashi sliding feet, the Japanese walked on an almost Suriashi but not a total one. It was taught to avoid showing the back of the sandals, which was considered poor etiquette, so Samurai would suppress the backward kicking as much as possible, and land from the tip of the foot instead of the heels.

In the Samurai ritual, only the joints below the hip joint are used for space movement and rotation. For example, when looking back, the body rotates using the legs, and the trunk never twists. The movement from sitting to standing (and vice versa) is the same, and in the Samurai ritual, the upper body remains vertical and does not move. Also, both hands remain on the thighs. This coincides with the teachings of Kanai Sensei, who stressed the need to have Toitsutai (integrated body). Next time you see a "classical" Japanese Samurai movie, you should be able to notice more of these features of the high-ranking Samurais. If you pay attention to the norms established by Ogasawara-ryu school, watching Samurai movies becomes a place to observe how the Japanese Samurai used to walk.

When you take the first step in a stationary position on your toes, the center of gravity moves forward, so you naturally kick backwards. Then, the upper body comes out in front of the landing foot only by inertial force from kicking. Since the center of gravity is still in front of the foot, the next foot naturally comes out in front of the landing foot. (*This YouTube video shows the concept in motion: https://www.youtube.com/watch?v=BLZfFmRNCJc*)

This propulsion principle is similar to Western-style walking. However, unlike Western-style walking, it is not necessary to kick out step by step, and after the first step, you can move forward with almost only inertial force. If you wear Ashinaka sandals, you will find that it is a footwear that realizes very energy-efficient walking.

Most people would find it interesting to know that Japanese clothing had much to do with the way Japanese walked and lived. Or, maybe it is the other way around. If you look at some instructional videos of how to wear **Kimonos** (traditional clothing 着物), it becomes clear that Kimonos create some conditions for walking and behaving. Japanese basically use a single T-shaped outfit, which is tightly secured with a sash called an **Obi** (belt 帯). During earlier times in Japan, many activities were traditionally performed seated on the floor. Because of this, the T-shape wrap-style Kimono allowed for easy movement but required attention to body movements so that one did not disarrange the Kimono. Wearing a single one-piece Kimono secured by one Obi, remembering that with a Kimono that was not tailor made and did not have shoulder pads or other supporting features, was a genius design but also required discipline to keep it tidy and elegant. Too much torsion of the waist, right and left, would have unsecured the balance of the Kimono.

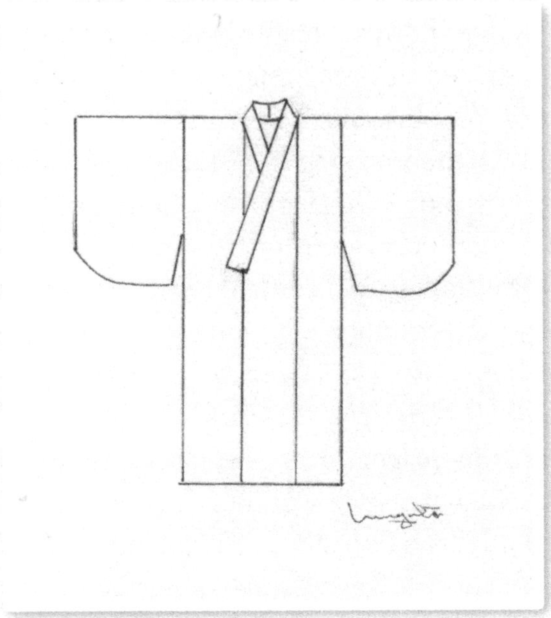

It is easy to visualize oneself with a bathrobe going haywire if you don't follow the Ogasawara protocol and make certain big moves. The key here is how much torsion there is at the waist and that is why the **Hakama** (pleated pants 袴) became necessary for riding horses or for more aggressive combat situations. The Aikido **Keikogi** (training outfit 稽古衣) are less of a problem since it is a two-piece garment with jacket and pants that allows more tolerance for major movements. The pair of pants makes the difference for stability as well.

The history of the Hakama for men dates back to the Kofun

period. (The Kofun period is often referred to as the period of about 400 years from after the mid 3rd century to the end of the 7th century.) Note the Hakama in the photo of ancient **Haniwa** (clay figurines 埴輪).

Hakama of the Kofun period were used only for men and took the form of thick trousers that went down to the ankles. The oldest Hakama in the Shosoin (Japan's historical storehouse) is said to be similar in shape to a Mompe (もんぺ), a term meaning a trouser-like garment for women that is believed to have its origin in the farming clothes of the Tohoku region. In the Heian period (794-1185), noble men began to wear them as formal attire, and they changed from practical to ceremonial. In the Muromachi period (1336-1573), when the culture of the aristocrats mingled with that of the Samurai in the Kamakura period (1192-1333), various types of Hakama were created for different purposes, but today, the split-crotch "horse-riding hakama" is worn, especially for martial arts like Aikido, Kendo, and Kyudo. Interestingly, Hakamas for women continue to be an important part of formal attire and you still see many female college graduates wear them at graduation time.

When Samurais walked, they carried **Daishoh** (two swords大小), literally meaning "big-little"—a Japanese term for a matched pair of traditionally made Japanese swords (**Nihonto** 日本刀) worn by the Samurai class in feudal Japan. And as we have seen, the characteristic of the Kimono wearing, on top of the carrying of two relatively heavy swords on the left side of the Obi, required that the hip/waist would not be rotating right and left in order to maintain the Kimono properly. Having the arms swinging and shoulders moving alternatingly front and back would have also disarranged the Kimono as well and was not proper under the guidance of the Ogasawara school etiquette. Hakamas helped to keep the swords more stable but still required a balanced walk to avoid the swords getting out of place or the Kimono belt becoming loose. In short, you could say that Kimonos require that one keeps the posture correct. Furthermore, moving long distances with armor required that Samurais walk efficiently to avoid major fatigue before getting to the battleground.

For reference, let's talk about the Western traditions. The typical walking style from Europe was that you would balance the body movement by walking with opposite extremities forward. You could observe this especially in military marches. That is, when you move your left foot forward, you would be swinging the right arm forward. This opposite lower half and upper half movement of the body, in a way, achieves some balance but this small torsion of the waist in opposite directions would become a major burden to the body for longer marches. But this style of walk or marching was introduced to Japan in the late 1800s as part of the Japanese modernization of the military. Also, Japan introduced western style calisthenics to the population in general as Japan introduced a military conscription system when the country was moving away from the Samurai class military system.

In the West, there has been a custom of marching in formation and in step as far back or even earlier than Roman times, and Napoleon's army is known to have taken this march to perfection. In Japan, the Samurais sometimes moved in corps, but it was not until the Meiji period (1868-1912) that the custom of lining up in formation and keeping pace with one another began to be practiced.

Although there are no paintings of Samurai moving in military group formations, there are prints of the travel processions for Sankin Kohtai (参勤交代) during the Edo period (1603-1867), and it is clear that they were not always in step with each other. (In principle, Sankin

Kohtai was the policy of the Shogunate that feudal lords were obliged to alternate living in Edo and their domains every other year. In addition, the feudal lord's wife and heir's son were required to reside in Edo on a permanent basis.) In Japan, "some" folk dances of the present day have in-step moves with each other, but traditionally there are few examples of perfect group in-step actions such as River Dancers from Ireland or the Rockettes dancers of Radio City.

Western clothing and footwear were also introduced and the wearing of the swords was prohibited, as many may have seen in the "Last Samurai" movie. With the westernization of lifestyle, many of the traditional Japanese traditions gradually started to fade away.

We do not have films or videos to see how Japanese people walked during the Edo period or before. We have to rely on visuals from paintings, wood block prints, some early photography, or the descriptions of foreigners writing about the way Japanese walked. However, there are some historical facts that draw our interest as to how people must have walked or run. Some are related to the speed of

Samurai troop movements in the 1500s, especially when you consider the scale of the troops, the difficulty of feeding and logistics, the weight of the armors, and the speedy deployment.

Others may relate to the Japanese messenger service, public and private, called **Hikyaku** (literally meaning flying legs 飛脚), as shown in the photo. We know the service had plans to deliver mail from Edo (old name of Tokyo) and other major cities.

Commentary of foreigners living in Japan in the early Meiji period observing the **Rikisha** (runners pulling light two-wheeled hooded vehicle 力車) and some early photographs give us some hints of the possible walking methods. In the last 50 years or so, many sports enthusiasts started to talk about **Namba Aruki** (ナンバ歩き) and is becoming a type of a health craze.

Namba walking is said to be a way of walking in which the right hand and the right foot are simultaneously produced when walking. Namba walking is characterized by the fact that the axis of the body does not sway easily. I tend to believe that instead of walking with the arms swinging, the upper body was relatively fixed looking forward, without arms swinging.

Let's cover Namba Aruki and the differences with Aikido. Many people connect Namba Aruki with Aikido but I believe it is too early to judge this way. As mentioned in the beginning of the chapter, Aikido moves synchronize the right and left sides of the body to effect techniques. The definition of Namba walk is similar, but Namba is targeted for walking or running energy efficiently and there are no major swing moves in the arms. The role of the arm swings becomes very different. Also, Namba Aruki is basically for straight movements while Aikido can be rotational and include steps back as Tenshin movements. In addition, the Aikido practitioner definitely needs to extend the arms to execute techniques. I will continue to explore the connections with Namba Aruki (walk) and Aikido as it is an interesting topic. One thing we could definitely say is that Namba Aruki was responsible for Samurai troop deployments and the Hikyaku messengers.

08 The Knee: Hidden Powerful Regulator

Historically, Japanese people did not use chairs and had a habit of sitting on the floor, sleeping on futons on the floor, eating all meals at tables on the floor (tatami), and working or studying at tables on the floor, a custom that dramatically changed after World War II. This means that the legs and knees of the Japanese were very well trained doing stand ups and sit downs many times every day. In addition, traditional toilets required Japanese to squat, which also promoted stronger leg muscles and formidable knee power. This is in great contrast to Westerners who used (and still use) chairs to sit, eat, work, and study, slept on high rising beds, and had toilets where one could sit comfortably as well.

Westerners know that whether it is a tea ceremony, Japanese Shodo calligraphy, or Zen meditation, all will require some severe usage of the knees. And Aikido people know very well that the **Hiza** (knee 膝) plays a big part in Aikido training. The Suwari Waza or Zagi (sitting technique 座り技 or 座技) is a requirement for advanced testing in Aikido. Let me add a couple of more Japanese words: **Za-I** (sitting position 座位) and its counter word **Ritsu-I** (standing position 立位).

An interesting point is that practitioners of Japanese martial arts most probably will assume that Seiza (squatting straight [also correct sitting] 正座 or 坐) was the traditional way of sitting in Japan by the Samurai. However, historical drawings and wooden statues illustrate a diverse reality. Under the rule of first Shogun Tokugawa Ieyasu, the country was unified (1615) and peace was maintained for about 265 years under the control of the Shogunate. It can be said that during this period of peace and tranquility under the first Shogun, a more courtly etiquette and sitting style was developed. History books say that sitting in Seiza became more common during the time of Tokugawa Iemitsu (1604-1651), the third Shogun. Under the third Shogun, a sitting style developed that was less formal and more based on combat preparedness; Seiza became the norm.

There is a theory that the current Seiza posture was adopted by both men and women in the Edo period, when the before-mentioned Ogasawara school was introduced into the manners of Samurai families. This is not to say that Seiza did not exist before, as Buddhist statues from the Nara period (710-794) and writings of missionaries from the Muromachi period (1336-1573) also depict that the same sitting style as Seiza was common among ordinary people before the Edo period.

The name "Seiza" is thought to have been coined after the Meiji era. Sei-Za is a combination of two characters meaning correct (正) and sitting (座 or 坐). While we can assume that Seiza custom was formalized about 400 years ago, the fact remains that even earlier than this, the Japanese traditionally spent a lot of time on the floor or tatami, repeating the action to get up or sit down, using the lower limbs efficiently for daily life.

Until the third Shogun Tokugawa Iemitsu (1604-1651), the formal way to sit in official costume was not to sit in Seiza, but to sit

in **Raku-za** (worshipful legs or nobleman's seat 楽坐). In the Raku-za sitting posture, the buttocks are on the floor, the soles of the feet are together, the hip joints spread apart, and the knees are dropped to the floor. The person sits with a straight back, as illustrated in the painting of Emperor Montoku (reign 850-858). The slightly looser sitting style is called **Anza** (安坐) or **Agura** (胡

座), perhaps closer to the so-called crossed legs.

The Raku-za sitting posture is basically the Bhadrasana posture in Yoga. According to Yoga instructions, this asana helps to strengthen the knee and hip joints, thereby reducing pain in these joints. Thus, it enhances the flexibility of the legs. Apparently, this asana helps the spine by releasing tension from the spinal region. According to the Yoga Institute, this asana helps to loosen the joints by flexing and stretching the tendons. The muscles of the pelvis, knees, and ankles become more supple.

Additional benefits of the Bhadrasana posture (summarized from

Bhadrasana Posture

Oda Nobunaga

Tokugawa Ieyasu

https://theyogainstitute.org/how-to-do-bhadrasana) are as follows.

1. Aids extreme stretching of both the superficial and deep muscles of the inner side of thigh. The interior of pelvis, groin, and reproductive organs are benefitted.
2. It also reacts favorably on the muscles and ligaments of the urogenital region. Also, facilitates supply of fresh oxygenated blood to the groin region.
3. Relieves tension in the sacral and coccygeal regions of the spine.
4. Practice in moderation and it provides relief in cases of sciatic, varicose vein, and menstrual disorders.
5. Improves posture and concentration, as the mind remains focused and relaxed. Great posture for long hours of meditation and reduces mental activity.

It is incredible to note that nobles and feudal lords with advanced ages managed to keep such flexibility of the knee and limbs. O Sensei incorporated this stretching exercise as part of his warm up as well, as we shall see later.

Tatehiza (立膝) is a "battle-ready" sitting position even when wearing full armor. It translates to "raised knee." Tatehiza was used by Samurai in many situations, but the most common one was when traveling. You could rest—even sleep—in Tatehiza, but still be able to move instantly to counter an attack. From Tatehiza, one could rise up, knee-walk (Shikkoh), or roll in any direction, all while drawing the sword.

I find it's interesting that many scrolls and paintings which depict historically important women also are seen using Tatehiza sitting posture. This would be seen as an unacceptable sitting form for women in today's etiquette, unless they are training in Iaido. The portrait is of Kicho or Nohime (1535-1612), the wife of Oda Nobunaga.

Aikido was created in the early part of the last century and even though it has a strong connection to sword techniques, the Aikido sitting techniques start from Seiza posture and not from Tatehiza. The Suwari Waza or Zagi techniques originated from combat training Katas (forms 型), which was in preparation to counter attacks inside the palace or residence, as mostly likely people would have been sitting in Seiza position when they were attacked.

It is a very unlikely situation outside of Japan, but Aikikai's test requirement includes these rather difficult moves. The explanation, when given, is that this will strengthen your inner muscles more than doing standing techniques and it will be great for training. The logic goes, if you can do the techniques sitting correctly, it would become much easier if you are standing up.

Suwari Waza leads to Shikkoh and requires moving on the floor with foot and toes in a vertical position. There are benefits in that we strengthen the connection between the big toe and the section in the brain that controls the balance. Posture also becomes a crucial part in doing Suwari Waza. Being able to be in Za-I position allows one to do proper **Zarei** (sitting bow 座礼) to the **Kamiza** (dojo front 上座), **Sensei** (teacher 先生), and also to your **Keiko Aite** (training partner 稽古相手), or if a little bit more formal your **Do-Yuh** (friend of the path 道友).

Etiquette Before and After Seiza

Since Zarei (sitting bow) is performed while seated in Seiza, it is necessary to know the etiquette before and after Seiza. In all martial arts, when sitting in Seiza, the left knee should go down first. When standing from Seiza, the opposite is true: the right knee should stand up first.

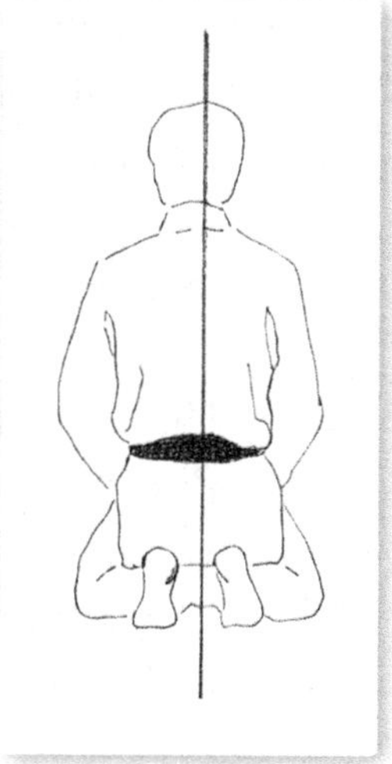

In the movements before and after Seiza, before placing both knees on the tatami mats or the floor, one assumes a posture called "**Kiza**" (騎座), as shown in the illustration. Kiza means that the buttocks are resting on the heels, rather than immediately sitting down in Seiza. When standing or sitting, one moves from one posture to the next in a flowing manner. It is also important to avoid touching the floor or tatami with the hands at this time.

The reason for sitting with the left foot first in Seiza is derived from the custom of Samurai in the past, who carried their **Katana** (刀) and **Wakizashi** (脇差) swords at their left hips, to maintain as much as possible a position in which they could draw their swords at any time.

Although there are differences depending on the type and style of martial arts, the general procedure for Zarei is as follows.

How to Sit in Zarei

First, straighten your posture. Starting with your left hand, place both hands on the floor in front of your knees, and touch the fingertips of both hands to the floor. When placing the palms of your hands on the floor, pull your fingertips inward so that they form the Chinese character for "eight" as in the following character: 八.

Next, bend your upper body toward the floor and assume the bowing posture. At this point, start bowing without looking at the person in front of you in the eyes straight ahead, and as you lean your straight upper body (no stooping) toward the floor, be aware that your gaze should naturally look downward but still maintain a sense of awareness of your surroundings.

When the bow is over and the upper body is raised, gently release the right hand and then the left hand in that order and place both hands back on the lap.

The benefits of Seiza are described in the West as follows:

1. Helps open up the ankle and knee joints
2. Puts pressure on the core muscles, engages them, trains them, and makes them stronger
3. Helps stretch your muscles
4. Helps correct your posture and strengthen your spinal cord

Seiza stabilizes the position of the pelvis so that the spine extending from the pelvis can also create the correct S-shaped curve. The advantage of Seiza is that it improves posture and straightens the spine. By flexing the right and left thighs, the inner iliopsoas muscles are engaged and the back muscles are lengthened, which leads to improved posture. With all these benefits, proper Seiza training has to be seen as an exercise to improve one's balance.

If you observe Kendo or Sumo matches, when the wrestlers or the athletes face each other to start the match, they will squat at each other's starting line to show respect in an action called **Sonkyo** (bow and show respect 蹲踞). This posture has basically the same meaning as Aikido Zarei, showing respect to the partner, but rather than fully squatting down, it is done squatting on the toes and sitting on the heels.

The importance of the knees is not limited to Suwari Waza or Shikkoh. Many senior Aikido instructors have stated that the knee plays an important role in Aikido. But few have been very specific as to how to use the knee to improve Aikido techniques. If, as they say, "the secret of martial arts lies in the knees," let's try to find out what those secrets are.

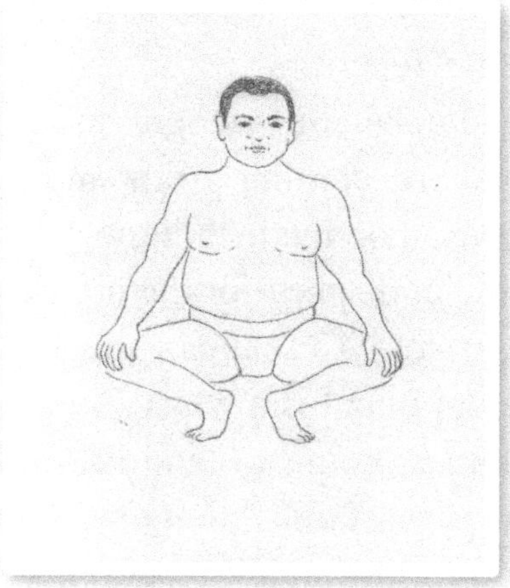

In the Aikido steps we spoke about, when you take the first step in a stationary position on your toes, the center of gravity moves forward and as long as you create an inertia leaning forward, you can continue to move quite effortlessly forward. In the earlier chapters about foot movements, we did not cover the role of the knee. However, if we add the knee element, it makes our body movements even more dynamic.

It doesn't have to be both knees at the same time, and you can

combine with a tilt of the body in one direction, flexing the level of firmness of the knee. As you can see from the illustrations included of some sports, people expect the ball to come in whatever "frontal" direction where the ball is and you need to be ready to fly to the left or right or perhaps dive to the ground or even jump higher. To do this, your natural physical preparedness of all your joints have to be ready, not locked standing straight, so that you can shift your body in whatever direction you need to go to get the ball. In Aikido, as we are based on martial principles, we need to expect attacks coming from all surrounding directions, so we need to stand straight but with knees flexed enough to move in "whatever 360 degrees" direction.

Most Aikido training is based on the repetition of prearranged attacks and countermeasures to those attacks. So we get used to anticipating where and how the attacks are coming from and we prepare our reactionary move directions based on the technique we are expecting. The Uke often takes ample time and space for the attack so the Nage is ready to execute the counterattack. As you can see in the illustration from soccer, the intent of the opponent is always shrouded in secrecy. You never know where the ball is going to land or go. In these circumstances, you have no luxury to stand tall with locked knees. The knees become the crucial element for quick reactions. We could summarize what flexible standing with knees slightly bent can do. As you recall, based on the Ogasawara School, teachings gave guidance as "Knees are not always fully stretched, but the hips are slightly lowered to eliminate vertical and horizontal movements."

If you use the body tilting and bend the knee in the direction you are tilting, you can make use of gravity to make certain moves:

- The first step starts quickly
- Weight transfer is fast and fine steps are possible
- Easy to "turn around"
- The arm swings and thrusts quickly, and the spring works
- Save energy
- Breathing becomes easier

The use of tilting, gravity, and weight shifts all combined with how you use the knee becomes a formidable power booster, with less muscle support in many of the body moves, helping you in all Tai Sabaki directional moves and more.

When Japanese people walk unconsciously on a daily basis, people say that they walk at the knee joint (center) while Westerners focus on the hip joint (base of the leg). This had a lot to do with the lifestyle of the Japanese and the wardrobe restrictions that did not allow longer strides when walking. As the lifestyle is dramatically getting westernized in Japan, the younger Japanese women now have less traditional baggage with the knee walk and are suddenly suffering from walking with Zouri sandals in Kimonos. With the fading of the Japanese tradition, it is also true that the Japanese are experiencing more knee injuries than they did before the Second World War. In Aikido, we see that many Japanese instructors who moved overseas started to experience knee problems as well. The aging of the Aikido population overseas also brings the problem of knee injuries as people have had stronger traditions of walking using their hip joints and had inherently less experience in fortifying the knee, muscles, tendons and nerves.

In order to avoid the risk of losing balance and falling, it is better that the position of the center of gravity is as low as possible, and the position of the body trunk and the center of movement are lower, administered at the knee joint rather than at the hip joint. All of this increases the sense of stability. We will cover more on this in the coming chapters.

If we look at specifics of how to use the knee, we see that it's possible to improve reaction speed in Aikido, reduce stress at the knees, and be able to execute more effective techniques without using too much muscle power. Also, by dropping your center, you can induce a rapid gravity drop in the center of the Uke and achieve a big throw.

The knee action in many martial arts is called **Hizanuki** (unplugging the knee 膝抜き) and can take many forms and combinations. In general, you "unplug" the knee, which can also be described as unlocking or allowing the knee to flex, and this lowers your center of gravity. It can be with both feet or slide out one foot and lower the knee. There is no torsion involved in this knee drop but the crucial element is that while you lower your center by bending the knee, the knee cap does not protrude and your center shifts forward.

If it is a simple drop, your center has to be more or less in the same position. Your knee cap should not shift too much in front, disrupting your own center. If you slip forward or backwards, try to maintain the knee cap behind the foot on the ground. You can apply Hizanuki in many grapple techniques to shift the center of your Uke and destabilize the Uke. This sudden center of gravity drop will definitely disrupt the center of the Uke (stealing your opponent's center), allowing the Nage to apply other techniques.

One point of consideration is to focus your attention not on the knee cap. When it comes to knees, most people are conscious of the front part of the plate, but what is needed to pull out the knee is actually the back of the knee. If you unplug your knees as if you loosen the back of your knees, you can drop your body with the center of gravity on your heels.

It is vital to keep the knees strong and flexible. Modern lifestyle has become very practical and we are supported by many tools to make life easy and comfortable. The downside of this is that we use our bodies less and less. But one must not forget that if we do not use some parts of the body, its functions will deteriorate.

I vividly remember Masando Sasaki Sensei who used to say "The Creator has made man very well. Man-made shoes wear out with use, but the more we use our feet, the stronger they become." Let's try to use our body parts wisely and use the legs, knees, and feet in ways that create more impact in our executions of the techniques and also keep us healthy.

09 Pelvis and Hip Joint with Tai Sabaki

English translation of the Japanese word **Koshi** (hip to pelvis 腰), which is used for **Koshinage** (hip throw 腰投げ), is very broad. It can mean waist, lower back, hip, and many others. The Japanese would roughly refer to the lower part of the spinal column to the pelvis. The character for Koshi is the combination of **Nikuzuki** (body flesh 月) and **Kaname** (pivot, pin, center, essence, substance 要). So, we can see that Koshi is an important area of the body.

They say Eskimos have many words to describe snow. If you study the Japanese language, you will also find many words that use Koshi as adjectives to describe not only the body parts but also moves, postures/stances, and even as an adjective to describe mental conditions as well. Here are just a few.

Koshi ga Hairu (settled Koshi 腰が入る) can mean that you are tackling the issue seriously.

Koshi ga Omoi (heavy Koshi 腰が重い) would mean that you are "slow to act," while **Koshi ga Karui** (light Koshi 腰が軽い) would mean "quick to act."

When used for negotiations, **Tsuyo Goshi** (strong Koshi 強腰) means firm or tough stance, and the opposite would be **Yowa Goshi** (weak Koshi 弱腰), meaning weak-kneed.

Koshi is the crossroad of many bones, muscles, tendons, ligaments, nerves, and veins connecting the lower half of the body with the upper half. It protects some organs and is crucial for the moves forward, backward, and balance of the body, while supporting the flexible and crucial spine and the upper body. Japanese Samurai trained hard to keep the Koshi section of the body strong, physically as well as psychologically. It is no accident that when the system of the Koshi disintegrates, as in the expression **Koshi Nuke** (absent Koshi 腰抜け),

it suddenly means "coward."

Oyobi Koshi (not fully engaged Koshi 及び腰) is another expression that means lack of nerve or indecisive attitude. In the West, people may use the word "spineless" for something similar. So maintaining the Koshi strength is a common theme whether you are talking about Sumo, Judo, Kendo, or Aikido. Even **Nihon Buyo** (Japanese traditional dance 日本舞踊) and Kabuki theater highlight the posture surrounding the Koshi. Koshi is crucial to maintain the fluidity of the body movements without losing balance. For the Japanese spectator, the Koshi represents a key point to focus when they see how good the Nihon Buyo dancers or martial artists perform.

Some people even say that Japan has a culture based on Koshi (hip to pelvis 腰) and Western culture is based on Kata (shoulder 肩). We can see the contrast in the preferred body style of males in the West and those representing the strongmen of Japan in the past. In Japan, this is highlighted in the difference of the "stable" pyramid triangle body shape versus the Western preference of inverted triangle body shape. The following illustration demonstrates what the Japanese considered to be strong (heavy around the Koshi). (*Image from Lyon Collection, Tsukioka Yoshitoshi [月岡芳年], artist 04/30/1839 – 06/09/1892, 力士梅ヶ谷藤太郎.*)

Compare this to the hand drawing of what Westerners view as strong (muscles around the upper body).

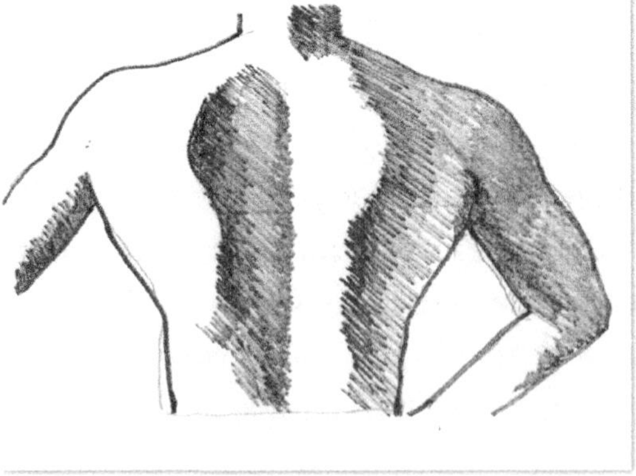

With the Western preference for the "inverted triangle" body shape, the emphasis is obvious: to strengthen the upper body muscles. Most superheroes in cartoons have exceptionally broad shoulders and muscles. We are somewhat indoctrinated to believe in the aesthetic beauty of these "inverted triangle" bodies as strong. This is similar to what we are taught to believe that high heels and pointed shoes are fashionable and good for women. As this book is not into fashion or aesthetics, let's continue to look at the roles and benefits of a strong Koshi.

We will touch base with Aikido but before getting there, first let's take a close look at one martial art that focuses a great deal in strengthening the Koshi. That art is **Sumo** (相撲). Sumo has very simple rules: force your opponent out of the ring or force your opponent to the canvas. The Sumo ring has a diameter of around 15 feet, and with bouts that sometimes last only a few seconds, it starts normally as a clash of two major body masses.

As heavy body mass multiplied by speed creates great force, Sumo wrestlers consciously try to gain weight to overpower the opponent. But what is not seen by the public in general is how much the wrestlers train to strengthen their Koshi and *flexibility*. They may look totally overweight but their flexibility is as good as many Yoga teachers.

Let me describe in more detail. When it comes to flexibility in Sumo, the first thing that comes to mind is "splitting," where you spread your legs 180 degrees and put your chest on the ground, because it is especially important to soften the area around your hip joints in Sumo. They say there are three main effects of creating a flexible hip joint by splitting the legs. The first is to prevent injuries, and the second is to create stability in the lower body, and the sense of "stability" increases the strength of the attack power. One of the reasons for the stability of the lower body is that the range of motion of the hip joint is expanded. This also allows extra range of motion in the muscles and joints, which makes it possible to respond instantly to the movement of the opponent. The third reason is that by performing a "split" with the pelvis upright, you will have a body that accumulates power in the lower abdomen (**Tanden**, which will be covered in later chapters).

If you watch Sumo closely, you will find that it, in essence, is a competition of who stays in balance and who loses it. Destabilizing the opponent becomes one of the major goals. If the center of gravity is high, it can easily collapse, but if the force is accumulated in the lower abdomen, the extra force of the upper body will naturally be released, and the body will have a sense of stability. As mentioned above, it is said that if there is flexibility around the hip joint, there is a sense of stability and it is easy to exert great power even in offense and defense other than standing. Wrestlers who have just entered the world of Sumo wrestling will be asked to create flexible hip joints by "splitting the Koshi." One of the exercises is **Mata Wari** (complete split 股割り) and the other is **Koshi Wari** (Sumo squat 腰割り).

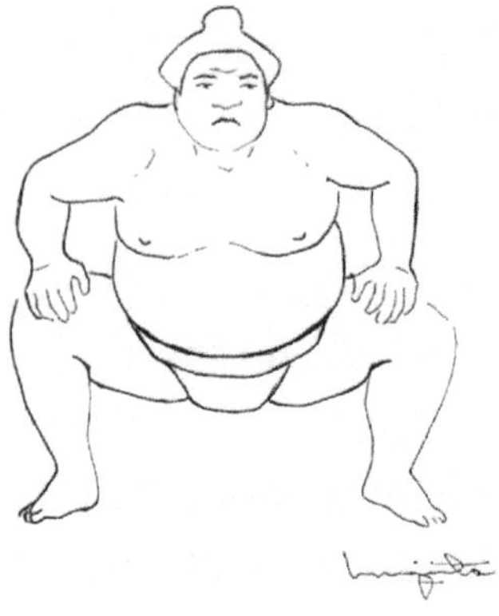

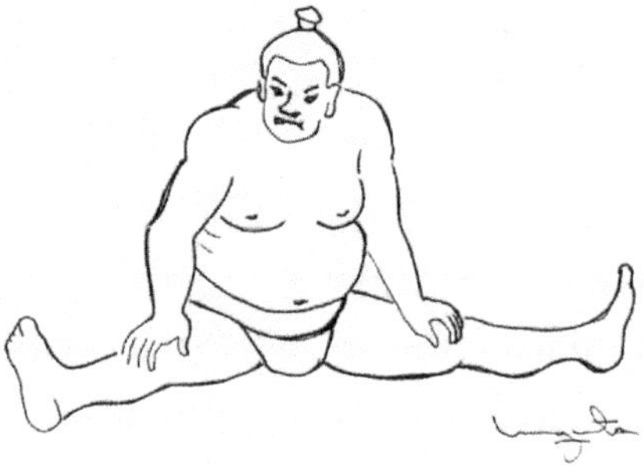

Another critical training for Sumo wrestlers is **Shiko wo Fumu** (Shiko stomping 四股を踏む), which strengthens the Tanden and the muscles of the lower half of the body, thereby cultivating stability and balance in the lower half of the body. It has the effect of a one-legged squat and helps one leg balance during kicking techniques. It also has the effect of softening the hip joints. As for muscles, it is best suited for training the quadriceps and biceps femoris (muscles on the back of the thighs) in particular.

In Japan, low back pain is called **Yo-tsu** (腰痛) with the characters meaning "Koshi" and "pain." Low back pain might be caused by various reasons but one thing that is certain is that the occurrence of low back pain would be minimal if you kept a daily dose of Koshi Wari, Mata Wari, and Shiko wo Fumu stomping as part of your health exercise regimen. We do not have evidence if there were people suffering from low back pain among the disciplined Samurai. I tend to believe that low back pain is a modern phenomenon in people with bad postures and/or lack of exercise. Samurai people, even their families, probably rarely had this problem because they sat, stood, and moved in a way that kept their spine in proper alignment following the Samurai lifestyle. Koshi Wari, Mata Wari, and Shiko wo Fumu stomping are some important exercises that strengthen the core muscles that support your back. A strong core is one of the best ways to maintain proper alignment, become agile, and attain strong balance, even in motion.

Not only do top-notch people have an attitude of **Koshi ga Hikui** (low Koshi 腰が低い), meaning modest and humble, but their actual Koshi positions are also inevitably low. A vertically high placed Koshi (imagine the military posture of standing at attention) may look good in our Western mindset, but in sports and martial arts, it is considered a contraindication. If you have a high Koshi when you are in a stance before performing an action, your body will be unstable and you will not be able to control it as you wish. We have seen a low Koshi in the illustrations of sports athletes that are waiting to react and move.

For that particular reason, if you want to move your body accurately, you must first move your Koshi first. If the rest of the body moves first and then the Koshi, the Koshi and the rest of the body will be blurred and the shape of the body will be distorted. Efficient movement is characterized by beauty and does not slacken, wave, or collapse.

If the Koshi is high, the rest of the body tends to waver more easily, and the Koshi and the rest of the body become separated, resulting in disjointed and uncoordinated movements, inaccurate and ambiguous movements, and lack of transmission of movements. If you want to move in a way that affects others, you must always lower your hips (connection with knees). If you lower your Koshi, you will inevitably find it easier to sit upright and maintain a more stable balance in your body. You can take a look at Sumo's Great Champions (Yokozuna 横綱) YouTube videos to see them in action.

A low Koshi stance has the added benefit of allowing the body to move without extra effort because the center of gravity is in the body and the body weight can be used to move the body. If the Koshi is too high, the body will not be able to "sit" on the Koshi and the body will lean. If the body leans, the body's center of gravity will be outside of the body, so there will be less energy to move the body using the weight, and the body will have to use muscle power to move. This will cause an unnecessary force to be put into the original movement, and the honest movement will disappear. With the slightest blur, the slightest tilt, the body collapses, and the force it is trying to transmit is lost. And because the force is lost, the body tries to move with force on its own, and the original form and movement are lost.

One word of caution. Advanced practitioners can put their bodies on their Koshi without dropping low, so they don't look like squatters.

Furthermore, even if they lower the Koshi, they will be moving so smoothly that you will not realize that their center of gravity has gone low.

We could talk more about other Japanese martial arts or cultural arts, but let's try to focus on Aikido. If you go to Hombu Dojo and attend Doshu or Dojocho class, you will find many of the warm-up exercises to stretch the hip joints. A case in point is the four major stretches of the Makko-ho Stretch Exercise. In fact, it is said that O Sensei met with the founder of Makko-ho, Nagai Wataru, in 1959. O Sensei was impressed with its value and immediately adopted the **Makko-ho** stretch exercises (真向法) currently used as warm-ups in Hombu Dojo.

Makko means correct posture facing straight, and the exercises to create that correct posture are called makko-ho gymnastics. It is characterized by straightening the spine and opening the hip joints. It is said to eliminate distortions of the spine and lumbar vertebrae, increase flexibility of the body, strengthen the lower limb muscles, and have a positive effect on the internal organs. Makko-ho is based on four simple bending exercises performed in a seated position. The cross-legged posture, the long sitting posture with both legs straight out, and the long sitting posture with both legs open as far as possible and ankles standing at an acute angle and the split-leg posture.

These exercises are to add flexibility to the hip joints, and as we have seen with the case of Sumo, the added flexibility is essential for the Koshi to move efficiently to generate power to throw the opponent.

Aikido has an additional exercise that is called **Funakogi Undo** (rowing exercise 船漕ぎ運動) to strengthen the Koshi and breathing timing. **Zagi Kokyu Ho** (seated exercise 座技呼吸法) is a breathing power training method in which the Nage and Uke are seated facing each other in seiza, and both of Nage's hands are held by the Uke. The objective is to learn to effectively exert the breathing power of the Nage and the aiki of the forearms. In addition to the forearms, it is also an exercise to learn how to effectively exert the power of the whole body. While the Funakogi Undo focuses on forward and backward motion, Zagi Kokyu Ho trains the Koshi in omni-directional motion.

A direct disciple of O Sensei, Okumura Shigenobu Sensei, used to teach beginner's class at Aikikai. I took his classes even after I joined the regular Doshu's class, as Okumura Sensei had many interesting anecdotes of O Sensei and was very knowledgeable about the history of Aikido and martial arts in general. Okumura Sensei taught how to swing with a **Bokken** (wooden sword 木剣). After the preparatory exercises, the next exercise was often a Bokken swing. But interestingly, I think his emphasis was not so much the precision of Bokken swings but to teach how to use the Koshi as part of the overall body movement.

First, we would hold the sword in **Migi Hanmi** (the right half of the body 右半身). Next, the right foot is retreated one step and the sword is raised. Next, the right foot is shifted in a clockwise direction (360 degrees divided by 28), and swing down the sword by stepping forward. Using your left foot as the needle of the compass, swing up while pulling back your right foot, and swing down while stepping forward.

Okumura Sensei would repeat this swinging motion of the sword 28 times, circling clockwise with himself at the center. During this practice, Okumura Sensei said two things to us.

One is: "Do not raise the sword with your hands/arms; swing it up with your Koshi. Do not bring the sword down with your hands/arms; swing it down with your Koshi." In fact, if you train hard doing **Suburi** (sword swing 素振り) and do it solely relying on your arms, one loses the power very quickly and gets exhausted. The arms should be an extension of the Koshi, based on the Tanden (which you'll read about in the next chapter) and swing down the arms integrated to the center, which is the Koshi. Okumura Sensei said, "It is the same when applying Aikido techniques. You must use the power of your Koshi instead of relying on the strength of your arms," he told us.

Another thing he said was, "It is not to cut the air. It is not to cut an invisible opponent. It is to cut yourself. You must cut away your own impurities. While doing Suburi, you are performing **Misogi** (purification 禊)."

This was Okumura Sensei trying to teach us O Sensei's way of **Masakatsu Agatsu** (True Victory is Self Victory 正勝吾勝).

A strong Koshi, in harmony with all the limbs and head/neck, is essential for a well-balanced motion in Aikido.

10 Tanden: Your Body Center

Since ancient times, **Tanden** (Tantien in Chinese 丹田) has been known in the Orient as a place to "knead the Ki (Qi)" of the human body. It was not just a health technique, but also a source of superhuman power. So, what is "Tanden"?

The Chinese character for Tanden includes "**Tan**" (medicine for long life 丹) and "**Den**" (a place/field to make it 田). In other words, Tanden is an important energy center for storing, enriching, and utilizing Ki as needed.

Today, however, few people know exactly what the Tanden is, and it is considered to be one of the lost pearls of wisdom of the body.

Now, in the 21st century, the benefits of yoga and zazen have been scientifically proven, and interest in Tanden has been growing again. However, most of the teachings are still limited to "training methods" for business people and athletes, or "relaxation methods" to relieve depression and insomnia. Of course, such approaches are effective, but they are only a small part of the functions of the Tanden!

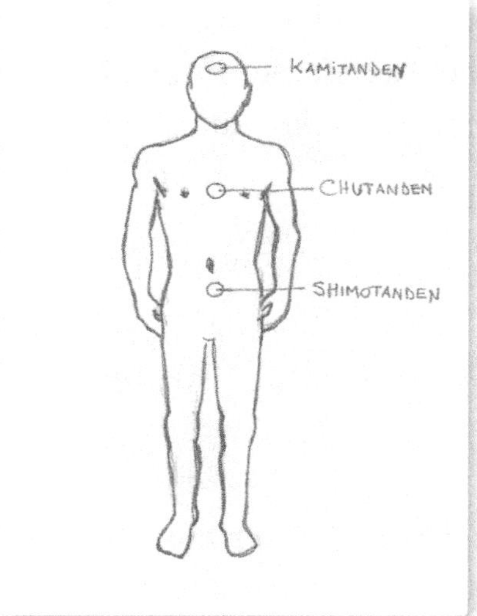

In **Taoism** (道教), which is the basis of Chinese thought and philosophy, the Tanden is taught as having three parts: the upper (**Kamitanden** 上丹田), middle (**Chutanden** 中丹田), and lower tandens (**Shimotanden** 下丹田).

The fundamental energy of the body is stored in the upper, middle, and lower tandens, respectively.

Upper Tanden: Shin 神 (character is God or Divine, but I would translate it as the center that controls the mind)

Middle Tanden: Ki 気 (life energy)

Lower Tanden: Sei 精 (essence, but I would translate it as vigor, vitality, even reproductive vigor)

It is interesting to note that of these, the upper Tanden refers to the pineal gland, which is often referred to as the "third eye."

The vital importance of energy in the body is underscored by this quote:

"Human life is a collection of qi. When they are gathered, they become life; when they are scattered, they become death."
— Zhuangzi (莊子), Chinese philosopher c. 369 BC – c. 286 BC

In addition, there is a generative relationship in which Sei/essence (reproductive vigor) generates the Ki/life energy, and from this Ki/life energy arises the Shin (center that controls the mind), and as such Shimotanden, which houses the root of the root, the (reproductive) vigor, becomes extremely important. It is also called **Seika Tanden** (sub-umbilical tanden 臍下丹田), because its location is below the navel. As such, Seika Tanden is the core of physical energy and is the same as Shimotanden. In other words, if the "lower tanden" (so-called "tanden" or "tanden below the umbilicus") is not full, then the fundamental place of the body is empty, which means that the body as a whole is weak. This is why Tanden is so important in martial arts, where the axis of the body is emphasized.

Where Is the Tanden?

Tanden has long been considered important in martial arts. If you are a practitioner of any martial arts such as Aikido, Karate, Judo, Kendo, Kyudo, or Shorinji Kempo, you may have been instructed by your Sensei that you will be more stable if the center of gravity of your body is in the Tanden area.

While the Taoist descriptions of Tanden include three areas as noted previously, there is no clear demarcation as to where the lower Tanden (Shimo Tanden aka Seika Tanden) are located. References talk about Tanden being a couple of inches below the belly button inside the abdomen, but there are no specific organs that represent the Tanden itself. But as is the case of Koshi, this area is the crossroads of the anatomical "ganglion," a network of nerves, which is called the "solar plexus" in Oriental medicine.

The solar plexus is where the autonomic nervous system is located and is sometimes referred to as the "second brain" or "abdominal brain" located in the abdomen. Activating the area where these nerves are gathered is called "training the Tanden" or "activating the Tanden," and the method has been explored in various ways since ancient times.

In Aikido, seniors have said that Funakogi Undo (船漕ぎ運動) or Zagi Kokyu Ho (座技呼吸法) and Kokyu Ho (呼吸法) techniques, in general, are effective to train and activate the Tanden.

You may see the connection of Kokyu-based techniques to training the Tanden. When it comes to training the Tanden, breathing exercises are often mentioned. A person's entire life begins with the breath at the moment of birth and ends with the death at the last breath. Breathing is a deep wisdom that touches the mechanism of life. It has been known since ancient times that this breathing is closely linked to the posture of the human body. Tanden sensation has been known as the physical sensation that appears between the breath and the posture of the body. While it is beyond the theme here, we all know how much O Sensei devoted his daily exercise to the concept of **Kotodama** (the spirit which is present in words 言霊) and Kokyu itself.

The most popular method of activating the Tanden is breathing exercises. It is said that breathing in combination with awareness of the Tanden (called Tanden breathing) can relax the body and mind, improve concentration and decision-making ability, and provide inspiration. The mechanism can be explained by the function of the autonomic nervous system. But then what is the autonomic nervous system? The autonomic nervous system is a nerve that works 24 hours a day to regulate the activities of the circulatory, digestive, and respiratory systems. There are two types of nerves: sympathetic nerves, which are active during the daytime and when the body is active, and parasympathetic nerves, which are active at night and when the body is at rest. When the autonomic nervous system is disturbed by an irregular lifestyle or stress, various disorders appear in the organs of the body.

In the Orient, it is known that the autonomic nervous system can be regulated by "abdominal breathing" with awareness of the Tanden. The solar plexus, the ganglion of nerves located in the sub-umbilical Tanden, is connected to the brainstem by a delicate network of nerves. The hypothalamus, a part of the brain located in the brainstem, governs the autonomic nervous system. When you breathe through your Tanden with abdominal breathing, nerve current pulses flow from the nerve network in your stomach to the brainstem. This activates the brainstem at once, and instinctive feelings of pleasure and peace are born. In other words... breathing with awareness of the Tanden can indirectly affect the autonomic nervous system! The stimulation from the Tanden is the access code to the brain stem. That's how it relaxes your mood and draws out a high level of concentration on a subconscious level!

If you master the art of Tanden breathing, you can deliberately create a state of mind called the "zone," where deep relaxation and high concentration coexist. The Zen monks and warriors of the past must have known this empirically. Recent scientific research has revealed that this was achieved through physical exercises such as Zen meditation and Sonkyo crouching.

The Sonkyo crouch (蹲踞) is a sitting posture in which the body is rounded into a crouch or in which the knees are folded and the hips are lowered to a standing kneel. It is also a ritual in Shinto. It is also one of the basic postures of Sumo. It is a posture in which one stands on tiptoe, sits deeply down, knees fully open, upper body upright, and center of gravity stabilized.

Another important method to strengthen your Tanden is to change how you walk. Instead of stepping forward and landing on your heel, you can adopt the Tanden walking method, which aligns more closely with traditional Japanese walking techniques. In Tanden walking, the entire sole of the supporting foot, which carries the body weight, pushes the ground backward instead of pulling up the front leg when moving forward. The back foot pushes the body forward while leaning slightly forward, with full awareness of the Tanden and sacrum, utilizing gravity. This way, the advancing foot lands on the ground with the whole sole or the toes. This method helps maintain a relatively level position of the Tanden. Walking in this manner increases stability and strengthens the forward push in martial arts. When movement is performed with awareness of the Tanden, even when executing Tenshin backward movements, it will be strong and stable.

Serotonin, the Happy Hormone

In the previous section, we looked at the mechanism by which the autonomic nervous system is regulated by the activation of the Tanden and the human body becomes healthy. One more thing I would like to focus on is serotonin. Serotonin, also known as the "happy hormone," is a neurotransmitter that has been attracting attention in recent years. When serotonin is secreted by Tanden breathing, aggression, impatience, anxiety, and fear are suppressed, and the mind and body become "refreshed."

Other benefits of Tanden breathing include reduction of fatigue, stabilization of hormonal balance, relief of insomnia, relief of pain, improvement of sensitivity to cold, improvement of metabolic function, increase in body temperature, and a sense of security, courage, and fulfillment.

As you can see, the benefits of activating the Tanden are many and varied, but one benefit you cannot ignore is that it improves posture! As we have been seeing, the posture (Shisei 姿勢) is crucial for balance and effective martial arts. Furthermore, Tanden breathing seems to secrete serotonin which activates muscles.

In fact, abdominal breathing, which senses the Tanden or the so-called crossroads of anatomical "ganglion," is not achieved by being aware of breathing alone. In order to regulate breathing, we must work on posture at the same time. In the human body, posture and breathing are one set. No matter how much we do zazen, if it is with a crooked spine, slouching posture, or stiff hip joints that are not fully extended, we will never be able to feel the Tanden.

Let me quote from the International House of Reiki (靈氣) and its explanation about "The Energetic System":

Within an energetic system are channels that distribute energy throughout your physical and energetic body. These channels do not hold a physical structure, as do blood vessels or bones. And, due to this lack of a physical structure, they are not usually seen by the casual observer. For this reason, it's important that you keep an open mind about these energy channels — they may be like the water within which the fish swim...essential to the life of the fish, but not seen or known by most.

Although these energy channels don't have a physical structure, they do have an energetic structure. Wherever the energy channels intersect, there are concentrated points of energy that either enable the open and fluid movement of energy along the different channels, or block that movement. Imagine the energy channels as highways and the concentrated energy points as roundabouts in a highway system. If either the highway or roundabout is congested, vehicles will either slow down or even stop moving.

If your energy channels or concentration points become congested, due to anger, worry, fear, or other inner/outer elements in your life, then you face the possibility of illness. Given that, it's of the utmost importance to keep these channels and points open, clear, and vibrant.

—Source:
https://ihreiki.com/blog/the_energetic_system/?v=7516fd43adaa

Only by following this virtuous cycle process can the human body correctly activate the application of the Tanden! We may say that the traditional costumes, lifestyle on the tatami mats on the floor, and strict code of etiquette of the Samurai helped nurture the benefits of this "posture adjustment" and beyond. If you follow the proper training of Aikido, consciously focusing on Tanden, the benefits will be limitless.

11 Critical Shoulder (Strength or Weakness)

In Budo, shoulders play an important role. We are instructed **not** to raise our **Kata** (shoulders 肩) while executing techniques. When your shoulders are tense, teachers may advise you to release tension from your shoulder muscles. There is a tricky Japanese expression that does not translate very well into other languages. **Chikara wo Nuku** (力を抜く) does not mean total relaxation of the muscles but it is just removing the stiffness of the muscles. So martial artists would advise you "Chikara wo nuke, ki wo nukuna" (力を抜け、気を抜くな), meaning "release tension of your muscles but do not let your mind/Ki wander." It is a small difference in nuance but as is in the case of meditation, we do not meditate to relax but rather to sharpen our senses. In the end, by sharpening our senses, improving our receptors, and improving our posture and breathing, as contradictory as it may sound, we achieve relaxation.

The shoulders can be strengthened by exercise but it is also a very delicate structure that can be damaged easily if you do not treat it carefully. It is probably the most freely movable part of the human body, with a spherical joint structure that allows for complex movements such as turning, pivoting, raising, and lowering the arm. The complexity of its structure allows this super mobility, and it's precisely because of that you need to know how to use the shoulders safely and wisely.

I am not a medical expert, so allow me to use the definition of the "shoulder" from the Wikipedia Japanese version.

The shoulder, when viewed as a motor organ, is known to be the most complex of the various joints in the human body. The term "shoulder" can refer (in the medical sense) to the shoulder complex (a complex consisting of the sternum, clavicle, scapula, ribs, and humerus), which is the shoulder in the broad sense, or (in the narrower sense) to the joint between the scapula and humerus (glenohumeral joint), which is part of the shoulder complex.

The shoulder complex is a complex consisting of the sternum, clavicle, scapula, ribs, and humerus, but the most important point is that the shoulder complex is a combination of the sliding scapula (on the back of the human body) and the glenohumeral joint located at the end of the scapula.

As a result, the shoulder as a locomotor organ can perform complex motions such as rotation (abduction, adduction), rotation (external rotation, internal rotation, upward rotation, downward rotation), elevation, downward control (pulling down), traction (forward traction, backward traction), flexion (joint flexion, horizontal flexion), and extension (joint extension, horizontal extension). It is also composed of many muscles and can produce strong force.

This phenomenal mobility can be achieved by the interconnected pieces working together but especially among them the scapula, also known as the shoulder blade (**Kenkokotsu** 肩甲骨) Since the scapula is only connected to the bone at one point on the clavicle, the rest of the bone is covered by or connected to various muscles. The scapula is connected to many muscles but importantly connected to muscles that are supposed to transmit and amplify the power of the lumbar abdomen. Therefore, by using the scapula, the muscles connected to it will be able to work. In other words, the muscles work differently depending on how you use your scapula. In order for the scapula to work well, it must be able to move as freely as possible.

To move freely means to make the muscles connected to it flexible, or in other words, not to let other muscles interfere with its movement. It is the scapula that supports the strong force of the arms, and that transfers the power of the lumbar abdomen to the arms for subtle control.

Here is the key. While every element forming the shoulder is important, it is crucial that the energy from the Tanden is smoothly relayed with as little obstruction as possible to the shoulder, then to the arms, and ultimately to the hands and fingers. You can see from the drawing the extra range of motion that scapulas provide to our arms.

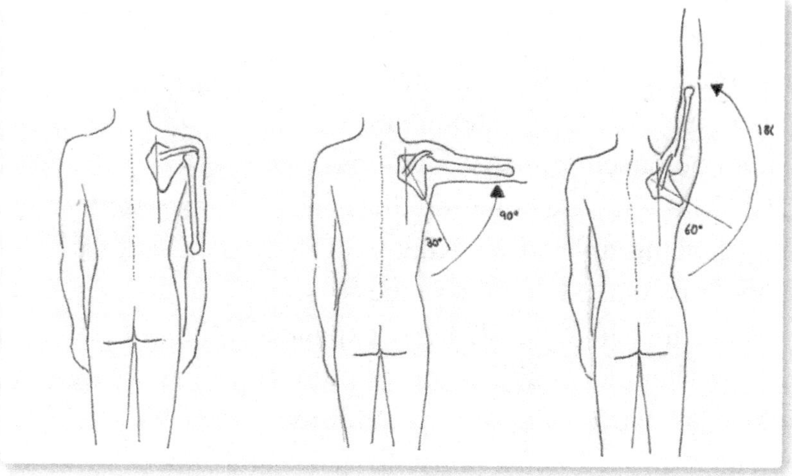

If you are sitting at a desk for long hours without using the shoulder blade, the muscles will lose their flexibility and restrict your range of motion, not only of the shoulder blade but the whole shoulder complex. This is evident if you look at people doing Suburi (sword swing 素振り), as described in the previous chapter, as those with a limited range of motion (stiff scapulas) will not be able to raise their arms up straight touching their ears.

To increase the range of motion of the shoulder, the time of elevation (described as when you glide the shoulders upward toward the ears, or "shrugging the shoulders") must be reduced and the muscles around the scapula must be able to move freely. The muscles supporting the scapula are also connected to the spine, ribs, upper arms, neck, and pelvis.

Thus, the scapula plays an important role in all parts of the body, not just for moving the arms back and forth and side to side.

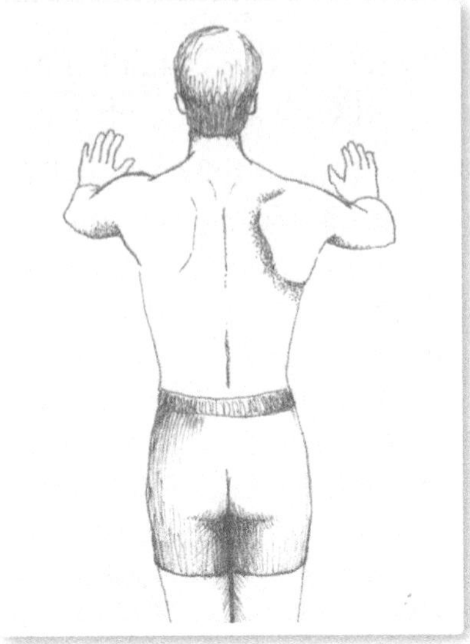

One way to train the scapula is the following: When you crawl on all fours, your shoulder blades are raised and look like wings on your shoulders. This condition is called **Rikko** (scapular winging 立甲). However, as we grow older and stop moving our bodies, the shoulder blades become stuck to the shoulders and cannot be raised properly.

The following are some of the benefits of being able to do Rikko. The scapula is relaxed and the range of motion of the shoulder blades is expanded. With a greater range of motion, the shoulder is able to move more freely, allowing for greater strength in all types of sports. In addition, when the shoulder blade movement improves, tense muscles relax and blood flow improves. This also helps to improve stiff shoulders.

The scapula is adjacent to the spine and ribs, which are responsible for stabilizing the trunk. In normal conditions, the scapula, spine, and ribs can move separately. However, for those unable to raise their scapulas as in Rikko (scapular winging), the shoulder blades (scapula), ribs, and spine get stuck to each other. Therefore, extra force is applied when the body tries to move, making the trunk unstable.

When the scapula can move separately from the ribs and spine, the trunk is stabilized and a beautiful posture can be maintained.

The shoulder girdle affects not only the shoulder area but also many other parts of the body. The abdominal muscles are one of them. If one can do Rikko well and the scapula moves freely, the scapula does not have to put its strength in an unnecessary place. The shoulder blades can then turn the extra strength to the abdominal muscles. When the abdominal muscles are strong, the lower body can move in a balanced manner. This will result in an even distribution of muscle to the legs and a beautiful, strong lower body that looks great. Remember that in Toitsutai, we want to use the 206 bones and around 650 skeletal muscles in a synchronized way to generate its maximum power potential.

The vulnerability of the shoulder can be seen in many of the Aikido hold techniques that use methods to neutralize the shoulder of the Uke. Whether it is 1 kyo, 2 kyo, 3 kyo, 4 kyo, Ushiro Ryote Dori, Morote Dori, or other techniques, the key controlling element is how to subdue the opponent by restricting the movement of the shoulder. Oftentimes, the Uke may push up the Nage's arm to restrict the movement of Uke's shoulder. Many of the other techniques may use other pivotal points and the shoulder might not be the principal restriction target, but the techniques include immobilization or restriction of the shoulder movement.

If the shoulder blade/scapula has a wide range of motion and you can use your abdominal muscles and fully utilize the lower body, you can generate enough power to disengage from the joint techniques applied to you and start a countermove.

Another way of looking at the beauty of Aikido is that by applying the techniques to Uke's shoulder, we are helping our partner to extend the flexibility of the shoulder blade. We recognize that the arms/shoulder would lose substantial power if the arm(s) are extended more than 45 degrees out of one's center. In Aikido we should try to keep our arms in a Kamae, as is shown in the photo, but always keep in mind that the shoulder range of motion allows us to stay "balanced in motion."

12 Utilization of Arms (Contraction vs Projection)

When I first started Aikido, more than half a century ago, I remember the shock and sensation I felt when I grabbed the wrists and arms of the seniors or teachers at Hombu Dojo. They were not trying to impress me but I could sense the "oneness" and firmness of their wrists and arms. I could feel the energy not coming just out of the arms. Was it Ki flow? Perhaps so, but now that I look back, it was definitely the "wholesomeness" of the arm and body in one.

Years and years of hard training have allowed me to build strong arms and the firmness that comes with the "wholesomeness" of the arm and body in one. Now I understand that the secret is not to treat the arm as a standalone limb. The arm has to always be connected with the Koshi/Tanden. As we have seen in the previous chapter, the shoulder also is crucial but only a pathway of the energy flowing from the Koshi/Tanden. Kanai Sensei used to say that rather than using partial muscles, it is better to use all the joints and muscles in the body to execute a technique. So the combined energy with all the body muscles had to move in sync in ways that countered the force coming from the limited use of the opponent.

If you compare Aikido to a fly-fishing rod, the fishing rod or lure alone can't do anything, but the angler's movements allow the thin rod and lure to move freely, allowing the angler to enjoy fishing in a mountain stream. Of course, Aikido is not fly fishing, but casting of the lure requires the expert movement of the angler. The casting is a result of a continuous movement of the angler to achieve its desired optimum result.

In describing the interconnectivity of the different parts of the body in order to execute techniques, Kanai Sensei hand-drew some stickmen to describe what he tried to convey. As Kanai Sensei's drawings below illustrate, the arm is not an independent limb but part of a wholesome part of the body movement.

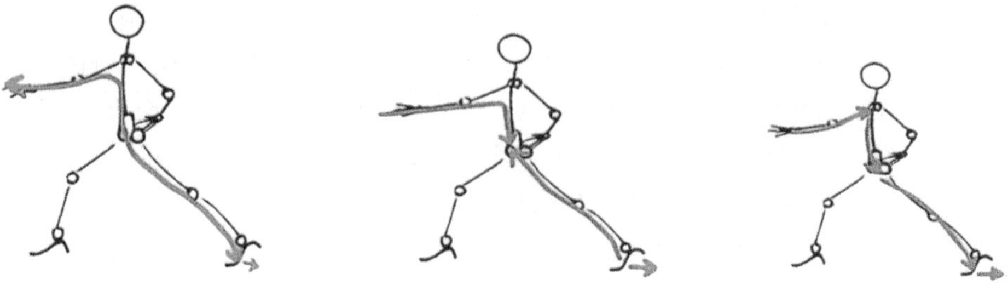

If we see the arm as part of a wholesome movement, many of the Aikido techniques become more powerful. Bear in mind that these drawings convey only two-dimensional energy flows and that Aikido—with Irimi, Tenkan, Tenshin, and Nenten movements—can bring many more dynamic variations. In order for the arm to be part of a wholesome movement, the arms have to be closely rooted, connected to the source of the generation of the power, as part of the Tanden, and strongly supported by the feet (see the position and angle of the rear foot), knee, and Koshi area.

One aspect of movement critical to integrating the power through the arms would be to start most power moves by keeping the armpits tighter to the cage of the body. In Japanese, we call it **Waki wo Shimeru** (tightening the armpits 脇を締める). Simply put, movement is necessary to link the muscles in the arm, back, and shoulder blade area. The human body is complex, and while we can move only the arms, we can also interlock the arms and shoulder blades. Westerners typically have stronger upper limbs and may not need to rely on tightening the armpits but for the 135-pound Kanai Sensei when he arrived in the USA in the 1960s, it became a necessity.

The scapula is said to interlock with the pelvis. By tightening the armpits, the serratus anterior muscle moves, the abdominal obliques move, and the diaphragm, transversus abdominis, and psoas major muscles can be used through the fascia. By tightening the armpits, the muscles known as the inner muscles can be used. As a result, the body can be used softly without stiffening the joints.

As you can see from Kanai Sensei's illustrations, the arms are not bent nor straight but extended like the warp of a Japanese sword. I always instruct on this point, as a bent or overstretched elbow cannot relay the power generated at the Tanden. Our arms can play an important role in executing Aikido techniques but we always have to be cognizant that as we use the opponent's arm as a leverage to subdue his or her movements, your own arm can easily become the tool for your Uke to do a **Kaeshi Waza** (counter technique 返し技) and use your own arm as a leverage for him/her. Leveraging the opponent's arm to execute techniques requires that the Nage's stance is always less prone to receiving Kaeshi Waza.

Kanai Sensei also said that most people get the illusion when thinking about muscles that contraction is the power solution. Imagine bulging biceps. But as Kanai Sensei's movements were a result of the integration of all body muscles, nerves, and tendons working in sync, he always mentioned that extending the force from the core of the body was more powerful. This is such a positive view of the use of integral power. Yes, contraction is also a force but think about the fly fishing example of casting. How are you going to achieve smooth casting with precise directions using contraction and big bicep movements?

Another important point is that blood flow within muscles fluctuates as they contract and relax. The ideal Aikido must allow a good circulation to allow the Ki energy to flow as well. Arms, as crucial as they are for implementing techniques, need to be the conduit of the Tanden power.

One case in point is when your Uke grabs your shoulder (Kata) for **Katatori** (shoulder grab 肩取り) and you wish to apply a 2 kyo technique. You either step back in Tenshin movement or step to the side to apply the 2 kyo technique. You do not contract in any case. You maintain the Maai (in-between distance 間合い) and apply a nice 2 kyo technique.

The arm is a major tool for Aikido. It can become a tremendous tool to destabilize the opponent, use it as leverage, work on it to subdue him. But all in all, using the arm requires great care to make it part of the extension of your Ki power.

13 Joint Techniques or Kansetsu Waza

Based on the technique classification by Jigoro Kano, the father of modern Judo, who studied and reviewed many of the **Kobudo** (Ancient Martial Arts 古武道) techniques, he grouped them as follows: **Nage Waza** (throwing techniques 投技), **Katame Waza** (restraining techniques 固め技), **Atemi Waza** (striking techniques including kicks 当身技), and finally **Kansetsu Waza** (joint techniques 関節技).

One of the prohibited techniques in Judo today for regular competition is Atemi Waza (当身技). Since Atemi Wazas were considered "techniques for killing," they were too dangerous to be used in Judo as a sport, but they were effective in serious fights on the battlefield. The term Atemi Waza is a general term for techniques such as striking, punching, or kicking at an opponent's vital points, and in some schools of Jujutsu, these "killing techniques" were also called "**Sakkatsu-jutsu** 殺活術" or "**Sappou** 殺法." While it is prohibited in regular competitions, Judo people train in katas that use the Atemi technique called **Kime no Kata** (極の形). It is a form that was developed in the arena of serious battles. According to the **Kodokan** (the headquarters of the worldwide Judo community 講道館) promotion tests, the Kime no Kata is included as part of the 5th dan black belt testing requirement but is not part of the regular competition as we know.

Judo also has Kansetsu Waza (joint techniques 関節技), which is the subject of this chapter. Joint techniques are techniques in which a fighter uses necks, legs, arms, knees, or other parts of the body to immobilize the opponent in Judo. Oftentimes, the fighter uses the opponent's elbow joint and applies a load in the opposite direction of the original direction in order to control the opponent. Joint techniques

are categorized in the same category as **Osae Waza** (pinning or suppression technique 抑え技) and **Shime Waza** (choking technique 絞め技), and include techniques such as **Ude Garami** 腕緘, **Ude Hishigi Jujigatame** 腕挫十字固, **Ude Hishigi Udegatame** 腕挫腕固, **Ude Hishigi Hiza Gatame** 腕挫膝固, and **Ashi-garami** 足緘.
— Source: *https://www.judo-ch.jp/english/dictionary/technique/*

Most joint techniques use the "principle of leverage" to restrain the opponent's joints. Because of the danger involved, the use of joint techniques is prohibited in Juvenile Regulations of Judo competition. In Aikido, too, many dojos introduce joint techniques much later in the youth curriculum to avoid damaging joints until the joints of the young practitioners are fully developed.

I highlighted Judo's Kansetsu Waza(s) in some detail as a comparison, as Kansetsu Waza(s) of Aikido are fundamentally different in application and end result from those in Judo. As Judo's are categorized in Osae Waza and Shime Waza and in most cases are for **Ne Waza** (groundwork techniques 寝技) restraining techniques. Judo has some similarities with wrestling restraining techniques while Aikido focuses on wrist, elbow, and shoulder restraints from standing and sitting movements. There are some interesting exceptions. When you look for old illustrations and photographs of Aikido by O Sensei, one finds some remnants of traditional martial techniques, as can be seen in O Sensei's Noma Dojo photos of 1936.

However, Aikido uses most of the Kansetsu Waza to destabilize the opponent and lead to throws and eventually enter into restraining positions. Aikido's destabilization leading to throws results in very dynamic projections which become the characteristics of Aikido. After destabilizing the opponent using joint techniques such as **Kotegaeshi** (flipping opponent's wrist 小手返し), **Juji-Garami** (cross elbow interaction 十字絡み), **Kosa Nage** (leveraging against the opponent's elbow 交差投げ), or **Tenbin-Nage** (leveraging against the opponent's elbow 天秤投げ), once the opponent has been thrown, the restraining techniques become quite different. Aikido uses leverages to immobilize the opponent by applying joint locks to the shoulder, as shown in the photos of pinning techniques, whether it is the techniques such as Ikkyo, Nikyo, Sankyo, Yonkyo, or Gokyo.

Even in Ikkyo, the angle of the arm is held at over 90 degrees to immobilize the opponent's shoulder. Aikido also maintains a high level of alertness to the surroundings, expecting the potential of additional attackers. Aikido techniques do not crouch over the opponent, but instead keep a good posture.

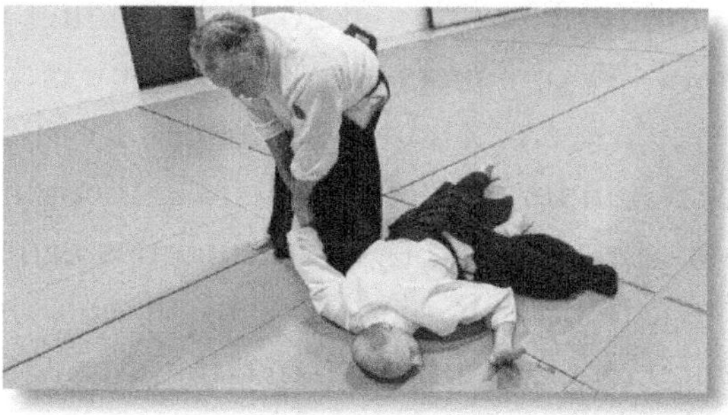

Judo is a martial art that begins with the opponents in Kumite, either Ai-yotsu stance or Kenka-yotsu stance, as seen in the illustration on the next page, whereas Aikido has a certain distance from the opponent, which is called **Iissoku itto no maai** (一足一刀の間合い) or "distance of one step, one sword." Basically, this is the interval in which one can strike the opponent by stepping forward and miss the opponent's strike by stepping back.

If you look at the **Kobujutsu** (ancient martial techniques 古武術) called Jujutsu and Taijutsu, which are the roots of modern martial arts as Aikido and Judo, you will notice that a large percentage of the techniques handed down are joint techniques and throwing techniques, and that striking techniques (such as punches and kicks) are rare.

When we see some YouTube videos of Aikido, many of the videos focus on techniques like **Tsuki Kotegaeshi** (突き小手返し). I think it is an effective and beautiful technique but personally I think Aikido is much more than just Tsuki Kotegaeshi. As the title of this book describes, I prefer to see Aikido as an art of maintaining your own balance while destabilizing the opponent, and there are many ways to do so. If you see the movies of O Sensei from 1935 organized by Asahi Newspaper, O Sensei conducts only one Kotegaeshi technique and another from Tachitori.

The goal of standing joint techniques in Aikido is to lock the opponent's joints one after another, starting from the joints closest to you, and finally break down the opponent's trunk and lower body stability. The original goal of standing joint techniques in Aikido is to hold the opponent immobile. This is because the techniques used in Aikido were originally intended to catch and hold an assailant, not to beat him in a match or to choke them until they pass out.

In addition, many techniques can be used only to apply force to the extent that pain is caused, and once the force is released, the opponent is unharmed, making it easy to subdue the opponent without doing harm. For this reason, standing joint techniques are also useful in self-defense and arrest techniques.

I think most Aikido practitioners must have had some experience of trying to demonstrate Aikido to curious friends, and the typical techniques used on those occasions are Kotegaeshi or Nikyo techniques. Even after gaining confidence in the effectiveness of the techniques in practice, it is often the case that when you do it outside the dojo with "strong" people, the techniques do not seem to work as well as when you applied it in the dojo. Rather than impressing your curious friends, you may end up making a fool of yourself.

The reason is simple. In the dojo, we train in movement; however, when you are a novice and demonstrating, it is most likely both persons are in static mode. Unless you are quite advanced in Aikido techniques, most techniques require some destabilization of the opponent to make the technique work. The title of this book, *Balance in Motion*, implies the Nage stays in balance even while in motion, "leading" the Uke to lose his or her balance somehow. It is advisable to apply the joint techniques when the Uke has lost some balance already. This is the key to successful application of joint techniques.

Interestingly, it doesn't take much to lead/move the Uke to lose his or her balance before applying joint techniques. From my personal experience, if you "extend" the arm, or shift the Tanden center of the Uke for a little more than one inch, it would make a big difference in the power of leveraging, as the grip of the opponent loses his/her optimum strength. The crucial element in using the leveraging is that you do not try to wrestle at the point of grip (fulcrum) but use your whole body, do the correct Tai Sabaki, reposition yourself, and find your optimum leveraging advantage and angle.

Muscles and bones act together to form levers. A lever is a rigid rod (usually a length of bone) that turns about a pivot (usually a joint). Levers can be used so that a small force can move a much bigger force. This is called mechanical advantage.

There are four parts to a lever: lever arm, pivot, effort, and load. In our bodies, this translates as follows:

- bones act as lever arms
- joints act as pivots

- muscles provide the effort forces to move loads
- load forces are often the weights of the body parts that are moved or forces needed to lift, push or pull things outside our bodies

One must always be cognizant that Kansetsu Waza works most efficiently when the total body works in sync. With good leveraging, the techniques work well even if there is a considerable difference in physique or strength. Even if you do not take big steps to increase the leverage, rotate your Koshi (hip), adjust your foot angle, and lower your shoulder, all of which bring one's energy closer to the Tanden. In this way, you can prepare your stance to achieve better joint techniques.

14 The Neck: Unprotected Vulnerable Part of the Body

In this chapter, we will try to highlight the importance of the neck. The neck plays a passive but important role in executing techniques or defending oneself. It is crucially important as it connects the body to the control center—the head.

Many of my seniors that trained Aikido before I joined Aikikai Hombu Dojo in 1969 told me they had trained in choke techniques (stranglehold) or neck-based techniques when O Sensei was alive.

In the book *Budo Training in Aikido* (*Original Aiki Jujutsu Densho* 合気柔術伝書, published April 1934), written by O Sensei, Ueshiba Morihei, published by Sugawara Martial Arts Institute (2001), you can see 16 technique illustrations against **Kubi Shime** (neck choke 首絞め) out of overall 166 techniques shown.

Astoundingly, almost 10% of the techniques shown are for counter attacks from Kubi Shime. Also, in the famous photos taken at Noma Dojo in 1936, O Sensei is demonstrating a technique called **Kubi Otoshi** (neck drop 首落とし). We can also observe a technique named **Kubi Osae Kokyu Nage** (neck holding Kokyu throw 首抑え呼吸投げ) from a photo from O Sensei's demo in Wakayama City in 1951. We still train in neck attack techniques such as **Ushiro Kubishime** (rear hold and choke 後ろ首絞め) or **Yokomen Uchi** (strike neck and temple area 横面打ち) techniques targeting the neck, but I have to say that the scope of the neck attack techniques has greatly diminished.

In many martial arts, there are "Kata," or forms, and in Karate in particular, which has many Kata, there are many movements that target the neck. Judo also has powerful Kubi Shime techniques and even Sumo has, as sports commentators call them, "vicious" **Nodowa** (a thrust at the opponent's throat 喉輪) techniques.

The reason why the neck is targeted is because it is the easiest place to target among the vital points of the human body. The human body has many vital points, including the brain and heart. However, the brain and heart are covered by bones such as the skull and ribs, as well as surrounding muscles, so it is difficult to damage them without a knife or blunt object.

In contrast, the neck, despite connecting important areas such as the brain and heart, has only a bone called the cervical vertebrae, and although large blood vessels such as the carotid artery and the laryngeal prominence (larynx) are widely exposed, there are no ribs or other covering bones, so even a bare-handed attack can inflict great damage on an opponent. For this reason, the neck is often targeted in the martial arts world. The cervical vertebrae in the neck, which support the heavy head, is the pathway for all the nerves that connect the brain to the body and is composed of seven bones. Therefore, poor neck posture is said to be one of the causes of various ailments, and it also affects the intestinal environment.

When we look at the role and function of the neck, the rotation angle of the neck triggers the movement of the crucial shoulders, hips, legs, and feet. As the head is the control center, the direction that the eyes move is followed by the neck and the body follows. As such, without a flexible and strong neck, the head and body would not work as one.

One immediate case in point is the front and back **Ukemis** (rolls 受け身), where it's vital to connect the neck with the abdominal muscles, extended arms, and legs for safe landings. A relatively well-trained neck is essential for good Ukemis. In other words, one could say that by repeating the training of Ukemi, one can strengthen a crucial connector between the head and body. Aikido Ukemis build up many muscles, but as far as the neck is concerned, the muscle build up will be with the front sternocleidomastoid muscle and the trapezius muscle that supports the sides and back of the neck.

The sternocleidomastoid muscle is the muscle that emerges from the front when the chin is pulled back. As the main purpose of the neck is to support the head to keep in balance, unless you build the muscles surrounding the neck like a bodybuilder, the neck can become a weak point in one's defense. In techniques such as Kokyu Nage (breath throw 呼吸投げ), **Irimi Nage** (entering throw 入身投げ), or **Sumi Otoshi** (corner drop 隅落し), applying force from under the chin pushing upwards and backwards can immediately result in destabilizing the stance of the Uke, very similar to the Sumo's Nodowa technique as you can see in O Sensei's photo of Sumi Otoshi technique, from photo collection at Noma Dojo. Note the Kanji in the image says Sumi Otoshi 隅落し.

In advanced techniques, some masters use the open palm to slide over the Uke's forehead backwards achieving the same result. It even can look like a fake Aikido technique, but if your neck shifts off your center backwards, regardless of being touched or not, the executioner of the technique has stolen Uke's center.

While it is not very common in today's Aikido training, it is interesting to see some photos of O Sensei doing Tenchi Nage (天地投げ), applying this force to the chin to unlock the neck and destabilize the opponent. It is quite different from the Tenchi Nage we train today. See the two pictures of O Sensei's techniques, also from Noma Dojo collection.

Let me add a few applications on using the opponent's neck to destabilize or subdue. When executing an Irimi Nage from the various attacks, including **Katate Tori** (single hand grab 片手取り), **Morote Tori** (grab with two hands 諸手取り), **Shomen Uchi** (front cutting attack to the head 正面打ち) and others, the Nage and Uke start the moves as two separate vertical axes.

As the technique progresses, the Nage gets closer to the Uke and can apply redirecting the shoulders of the Uke to destabilize and eventually bring the neck of the Uke next to the front of the Nage's chest. By locking the neck of the opponent on the chest, the Nage not only constrains the move of the opponent but achieves major destabilization of the Uke by tilting the axis of the Uke. By Nage holding the neck firmly in front of one's chest, the Uke has a hard time regaining balance.

In the Japanese language when you say **Kubi ga Mawaranai** (one cannot move the neck 首が回らない), it colloquially means to be in a quagmire of debt. So, not being able to move the neck means something extremely precarious. By training hard in Aikido, we have to avoid a situation where you can not move the neck.

Another utilization would be in the pinning position to lock the movement of the neck by gently adding a Tegatana on the neck. This would totally immobilize the Uke on the mat as can be seen in this rare photo of O Sensei.

15 Head, Hips, and Principles of Kuzushi

Perhaps for the reader, it may seem odd to read a chapter including the head and hip for Aikido. I decided to add it, as this was one of the topics Kanai Sensei used to talk about as being crucial elements for destabilization.

In Aikido, **Kuzushi** (destabilization 崩し) is the act of making the opponent's balance unstable while maintaining one's own stable stance. Kuzushi, while not highlighted very much in Aikido teachings, is a very important and essential part of the techniques in Aikido. By destabilizing the opponent, you can move or throw a large opponent with only a small amount of force.

The concept of Kuzushi is prevalent in most martial arts that focus on **Kumite** (grappling 組み手). Kumite (Kumikata) is to make an appropriate distance from the opponent (a distance slightly closer than arm's length, slightly bent at the elbow), then holds the opponent's front collar slightly below the collarbone (**Tsurite** 釣り手) with one hand and holds the outside of the opponent's elbow (puller's hand) (**Hikite** 引き手) with the other.

But as Aikido differs from other Kumite or grappling martial arts in how we start the moves, Aikido Kuzushi becomes somewhat unique. We will touch base in more detail but let's continue with Judo for the time being to see the contrast.

In Judo, the techniques of Kuzushi (destabilizing an opponent) are based on the use of Kumite, as we have seen in the joint techniques chapter. It is not only the use of hands and arms to destabilize an opponent, but also the use of movement, body weight, and whole-body strength to break down an opponent. In Kumite, both hands should be lightly gripped, elbows slightly bent, and sides lightly tightened. Judo rules state that the standard kumite in **Tachiwaza** (standing technique 立ち技) is to "grip the upper part of the opponent's right upper garment sash with the left hand and the upper part of the opponent's left upper garment sash with the right hand"; any other part of the grip is considered a foul or subject to a time limit (within five seconds).

The key to advantageous destabilization is to be aware of **Happo-Kuzushi** (八方崩し), or collapsing the opponent in eight directions: front, back, left, right, right front corner, left front corner, right back corner, and left back corner. The Japanese word Judo means "gentle way" in English, and its defining technical principle is to use the strength of an opponent against them. But all of these Kumite techniques require substantial force to achieve the Kuzushi of the opponent. It is for this reason that Judo has weight classes and men and women do not compete together.

Aikido doesn't have competition but was created based on a rich Japanese martial arts background. The attacks can come in a multitude of forms and sometimes with weapons. For regular training that uses hold/grab techniques, we have Katate Tori (one hand hold 片手取り), **Ryote Tori** (two hand hold 両手取り), Morote Tori (two hand on one arm 諸手取り), Katatori (shoulder grab 肩取り), **Ryo Kata Tori** (both shoulder grab 両肩取り), Ushiro Kubi Shime (choke hold from back 後ろ首絞め), **Ushiro Ryo Kata Tori** (both shoulder grab from back 後ろ両肩取り), **Sode Tori** (sleeve grab 袖取り), **Hagai Jime** (rear bear hug 羽交い締め).

With other empty hand attacks, we have Shomen Uchi (front cutting attack to the head 正面打ち), Yokomen Uchi (strike neck and temple area 横面打ち), and various level **Tsuki** (thrust 突き) attacks.

Furthermore, we have techniques to counter attacks using **Tanto** (dagger 短刀), **Tachi** or **Bokken** (wooden sword 太刀、木剣) and **Jo** (staff 杖). Aikido does not train against kicks but many of the moves are applicable to counter against kicks.

Training to defend and counter so many different types of attacks prepares us in Aikido to be very cognizant as to how to use every part of the body and body moves to destabilize the opponent against all those different sets of attacks. As the characters of Aikido (合気道) mean "the Way of Harmonizing Energy," it is essential that the Nage always should maintain strong balance and find ways to destabilize the opponent. The Nage should induce destabilization of the Uke (steal the center of the opponent) to execute the techniques. For the sake of good training, that destabilization has to be real. The only exception is when advanced people help beginners show the path of the moves to destabilize them. Choreographed moves create a false sense of good training and may be good for the egos but do not teach true destabilization of the opponent.

The start of the Kuzushi is in a well-timed and well-placed Tai Sabaki vis-a-vis the opponent. Oftentimes for training, we start the technique with a static position, both well anchored and stable. Even with static beginnings, it is essential that the Tai Sabaki action is taking the **Sen no Sen** (先の先) action. Kendo uses the term Sen no Sen, which means to strike at the moment when the opponent's mind is trying to strike us. Sen literally means "before," so in this case, you take action reading the mind of the opponent and go just before the opponent is trying to execute his or her attack.

There are many ways you could apply the spirit of "Sen no Sen." You do not let the Uke grab in his optimum position. You are already rotating your wrist before the Uke grabs you to maintain your center, under the control of your Tanden. You could be initiating the rotation of your hip and feet to locations that would be more defensible for you. You could use your hand that the opponent may try to grab as a bait.

Or, you could be initiating an Atemi strike. If and when the opponent already holds you tight, even good Aikido practitioners will have a harder time to move the opponent. So it only makes sense to be on the move already to counter the attack. Timing is important. Not too soon, not too late. Keep your core strong and be on the move.

In order to destabilize, Aikido does not use forward foot sweep (**Deashi Barai** 出足払い) like in Judo, nor do we grab the legs (except in **Aiki Otoshi**, pulling on the opponent from the side, grabbing the back of Uke's knee from the front, lifting Uke up, and throwing Uke backwards 合気落とし) to bring down the opponent like in wrestling.

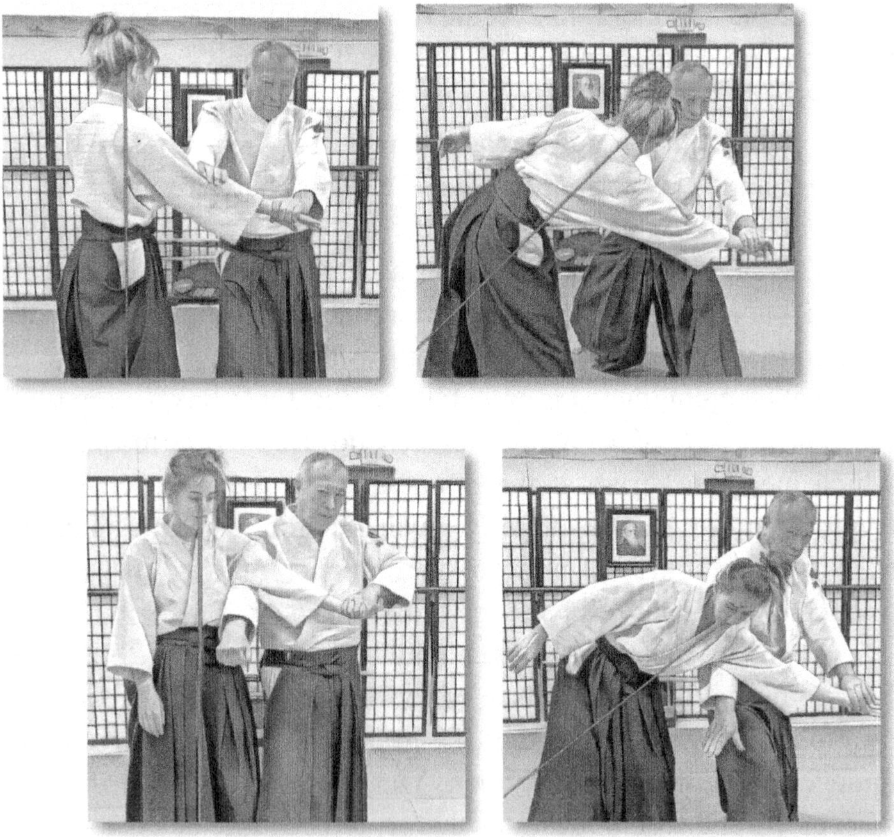

In Aikido, we often use the torsion direction difference of the upper body and lower body to destabilize *while moving*. We often use the inclination of the axis of the opponent's body to achieve the Kuzushi. We may use the wrist, arm, or shoulder of the opponent to basically generate conflicting directional energies of the upper body and the foundation of the lower body to destabilize, but they always have to be part of an integrated body move. In the photos, note that destabilizing occurs when the Uke's axis is no longer vertical.

Kanai Sensei often mentioned that the head and hip played a major role in destabilizing the opponent. He said that if the head and hip area, as two key weighted areas of the body, is out of alignment, people could easily lose balance.

Let's take a close look at the head, which is a new problem with people watching their cell phones too long. As more people stare at their cell phones tilting their head forward, people are starting to have major neck pain and beyond. The average human head weighs around 5 kg. or 11 lbs., and all that is balanced on just seven vertebrae in your neck and supported by around 20 muscles that are responsible for moving your head around and keeping that weight in place. By looking down at your phone around a 45° angle, the force your head puts on your neck increases all the way to 22 kg. or close to 50 lbs. of weight! It is a significant amount of weight to use it to your advantage.

Of course, Kanai Sensei was not thinking about cell phone problems, since he passed away before the iPhone was introduced. Nonetheless, he was very well aware of using the weight of the head to destabilize the stance of the opponent. Think about tilting or twisting the axis of the Uke, extending the arm or upper body away from the body center of gravity and adding the weight of the head tilting sideways. That is the magic of Aikido. Techniques like Kokyu Nage (breath technique throws 呼吸投げ) can only be achieved with this formula.

Distorting hip balance is naturally very crucial as well, as this is the section of the body that connects the lower foundation of the body with the upper body. Tilting sideways, backwards, or forward, or having the whole hip twisted, can be detrimental for the Uke to keep the balance intact. Leading the Uke to tilt the hip in any direction, with the extra weight of the head, can allow the Nage to destabilize the Uke much easier without too much physical effort.

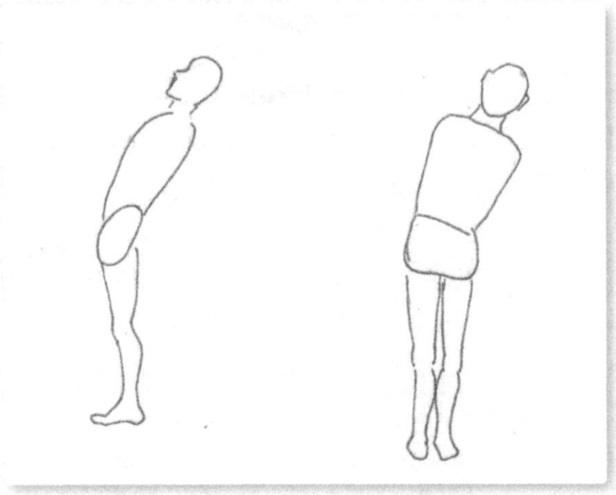

We can use the wrist joints, arms, elbow joints, and shoulders to accelerate the destabilization, but in order to achieve fluid and dynamic Aikido, one must use the two major weight masses of head and hip. Since most drawings or photos are two dimensional, it is difficult to appreciate how the destabilization of the body vertical axis takes place but one has to understand that it is not only the extremities that do the job. It is not that you have to grab the head or the hip but it is how to use them as tools to do the job.

Kanai Sensei executed many techniques unique to him using this method. His **Maki Otoshi** (rolling drop 巻き落とし), **Makikomi Otoshi** (rolling/wrapping drop 巻き込み落とし), and **Kata Hiki Otoshi** (pull shoulder drop 肩引き落とし) are a few of those examples of his technical interpretation. Kanai Sensei combined these techniques with his spiral knee drop to use the head and hip body mass weight to his advantage.

16 Redirecting Uke's Energy

One of the quotes from Ueshiba Morihei, founder of Aikido, says as follows:

Sometimes behave as air, sometimes like a stream of water, and leave the body in response to the opponent's movement.

ある時は空気のごとく、またある時は水の流れのごとく、相手の動きに応じて身体を預ける.

Aikido can be described in many ways, but one thing that highlights its uniqueness is that it is circular, spherical, and flowing. Some people call it the art of the flowing Ki energy. In Japanese, the term would be **Ki no Nagare** (気の流れ).

Ki no Nagare is a term in Japanese martial arts, including Aikido, that can be translated and understood as "the flow of energy" or "the flow of spirit." In the context of martial arts, particularly Aikido, it refers to the concept of moving with the natural flow of energy, both in oneself and in relation to an opponent.

Aikido's Tai Sabaki, in essence, is to avoid receiving the full attack impact from the opponent and, in most cases, it is used to redirect or unite with the energy of the attacker. Aikido is performed by blending with the motion of the attacker and redirecting the force of the attack rather than opposing it head-on.

Aikido's system of techniques generally involves moving in accordance with the opponent's movements, redirecting the flow of energy, and throwing when the opponent is thrown off balance. Let's use the example of a basic Aikido technique called **Ikkyo** (first teaching 一教). Ikkyo is, as the name implies, the first teaching and the

first technique that is generally taught. Technically, it is simply an elbow hold. However, it is not the first technique because it is simple or because it is easy for beginners to practice.

Ikkyo may be a simple technique, but has tremendous depth of redirecting the grab or frontal attack by manipulating the elbow and redirecting the flow of the attack. In Judo, Kuzushi involves grabbing the opponent's collar and pulling toward you to throw the opponent off balance, whereas in Aikido it involves manipulating the opponent's arm joints to throw them off balance. This is what makes Aikido different from Judo and Wrestling.

In Aikido, the elbow hold is not a movement to pull the opponent inward, but a movement to turn the opponent in the opposite direction, or in other words, to throw him or her far away. As you can see in the drawing, the Uke is attacking with Shomen Uchi and the Nage appears to be blocking with the right arm, but in fact the left arm is upending the attack arm and the right arm is only going to be used to crank down the attacking arm of the opponent. This is a clear case of redirecting and leveraging the attacker's energy.

Even though they are the same Kumite-type martial arts, the atmosphere of Judo and Wrestling are completely different from that of Aikido. In Judo and Wrestling, both practitioners lean forward and pull each other, whereas in Aikido, as the movement of Ikkyo typically shows, the practitioner does not pull the opponent toward him or her, but rather the majority of cases extends forward and throws the opponent far away.

Redirecting the energy coming upon one can be achieved with good Tai Sabaki but it doesn't have to be confrontational. If Aikido is known as the art of harmony (**Wago no Budo** 和合の武道), the redirection of the attacking energy also has to be done firmly and flowingly but not be collision-like. Rather than thinking forcefully to redirect, we might be able to see it as guiding away or leading away.

In his book *Aikido*, Ueshiba Kisshomaru Doshu compares Aikido to Karate moves and comments as follows:

In Aikido, although there are some thrusts and kicks, the movements are diverse, and the techniques can be seen to be concentrated in a completely circular motion, or in spherical motions.
— *Aikido*, by Ueshiba Kisshomaru, p 74, published by Kowado, 1957.

He continues his comment on Sabaki, which is the key mechanism of redirection.

Sabaki includes hand-sabaki, foot-sabaki, and body-sabaki, etc. If these are disjointed here, it cannot be called effective sabaki. Sabaki moves that take you front and back, left and right movements, all of these must be included in a coherent circle movement. The Sabaki emanating from the aiki Kamae stance extends to the fingertips, and the body moves in circles and spirals. When you look at a single movement in Aikido, Aikido is like a dance and ballet, and when the opponent comes, it engulfs the opponent in its spherical movement and

also projects the attacker away. It is just like a gear with an absolute center.
— *Aikido*, by Ueshiba Kisshomaru, pp 93, 94, published by Kowado, 1957.

In doing research for this book, I found the following statement from Emi Hirofumi, 6th dan, Dojocho of Shinbukan Dojo in Osaka, Japan. It follows:

My teacher Kadokawa Shigemi said, "If you make a circle bigger and bigger, infinitely bigger and bigger, it will eventually become a straight line. That is, the largest arc is a straight line. Aikido is a way of learning to eliminate waste. In other words, Aikido is about eliminating, perfecting, and sublimating into an efficient technique. This is what I have been taught.

Ultimately, it is important and our goal is to sublimate the techniques to the rational and efficient techniques as Kadokawa Shihan says, and to reach a movement that looks linear but has the principle of "Circle-rotation/Sphere-rotation" firmly in it. First of all, we believe that it is ideal to prevent the practice from becoming a practice of hand techniques and arm strength, as mentioned above, although there is a lot of waste, and to be able to embody the technique as an aiki movement in which the hands, koshi, and feet are in total agreement, or to understand it physically before entering the stage of sublimation of omission.

When we practice with this "slow, round, and large" awareness and with the flow of KI in mind, we can see that everyone is unknowingly aware of the median line, aware of the softness of the knees, and relaxed throughout the body, and then becomes aware of the correct aiki movements with hands, koshi, and feet in unison.

Let me also share Bruce Lee's most famous philosophical quote:

> ***Be Water, My Friend.***
> *Empty your mind.*
> *Be formless, shapeless, like water.*
> *You put water into a cup, it becomes the cup.*
> *You put water into a bottle, it becomes the bottle.*
> *You put it into a teapot, it becomes the teapot.*
> *Now water can flow or it can crash.*
> *Be water, my friend.*
> — Source: *https://brucelee.com*

17 Leveraging

With the multitude of attack techniques that Aikido has to handle, leveraging (**Teko no Genri** 梃子の原理) becomes an important tool to execute counter techniques. You could say that the great majority of Aikido techniques use leverage to a large extent. Leverage can be used for controlling techniques as well as dynamic throws. *Generally*, in the body, bones serve as levers (points of action), joints serve as axes of rotation (fulcrums), and muscle contractions serve as forces (points of force). Whether it is Kotegaeshi, Shiho Nage, Sumi Otoshi, Shomen Uchi Ikkyo, Nikyo hold, or Juji Garami, all use leverage.

But as Aikido often uses Flow Energy, many of the physical leverage movements can be integrated with great Tai Sabaki movements with minimum physical contact as well, sometimes even with none. Kokyu Nage techniques fall into this category. This makes Aikido movements very unique within the martial arts world.

But leverage, even as a mechanical operation, can be much more complex than just levers, fulcrums, and points of forces. It is obviously much more than two dimensional as it focuses on many moving parts in a very dynamic way. This makes it very difficult to learn how to use leverages, as very few instructors teach the mechanisms of dynamic and multi-dimensional leveraging principles. This is why Aikido takes so many years to learn. In essence, it is like learning the body anatomy of the opponent to affect techniques and control techniques. Many learn by trial and error, training over and over, but as the application variabilities are so many, unless someone deciphers and categorizes it well, it takes a long time to learn. For example, Kata Tori Nikyo is very different from Katate Tori Nikyo, and the list goes on as there are so many varieties.

Once you start to understand the principle of the leverage, the characteristics of the location and attack, and the applicability of the leverage technique, it becomes easier to understand. Suddenly how to use leverage to destabilize, apply pain, and restrain the opponent becomes clearer. Even if the circumstance and location are different, once the principles of using leverage are known, all becomes much easier.

Leveraging is a powerful tool to execute techniques but it cannot function until the opponent is off balance. A powerful leverage technique happens when the Nage does a good Tai Sabaki and repositions to a more advantageous angle, and then does Kuzushi (destabilization) of the opponent and applies the leverage techniques. How about the key elements to make the leverage work? How to line up the points of action, fulcrums, and points of force to effectively execute the leverage action?

Let's go one step further than simple levers, fulcrums, and points of forces. If we were to apply leverage techniques based on the following four points, perhaps we will start seeing with much more clarity the concept of effective leverage. The four points are as follows.

1. Anticipate the move of the opponent (**Sen no Sen** 先の先)
2. Fixation of contact point (**Setten Kotei** 接点固定)
3. Use of reactionary force of the opponent **(Handoryoku no Riyou** 反動力の利用)
4. Construction of leverage complexes (**Fukugo Teko no Kouchiku** 複合梃子の構築)

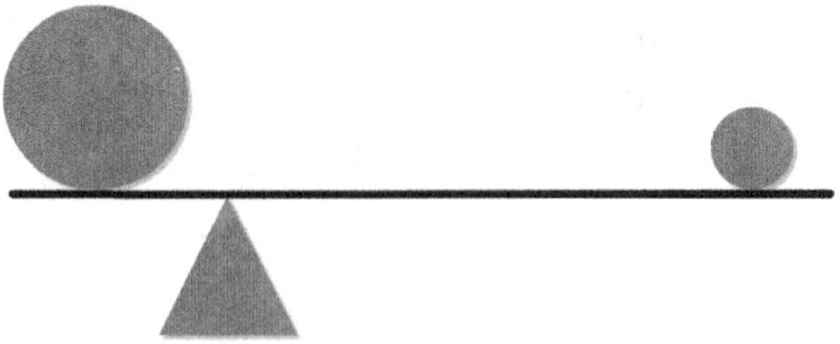

I do not wish to simplify the learning of Aikido, but once these four principles become a cognizant part of Aikido training, the improvements will be much faster.

Anticipate the Move of the Opponent

As the illustration shows, when someone does a Katadori attack, and the Nage fixes the hand grabbing the shoulder, the leverage technique is about to take place. This illustration depicts the Nikyo technique already in full effect. Aikido is often practiced from a static position and we know how difficult it is to execute the technique at this point if the Uke grabs you with all his or her might from a static start. It takes a major effort to move the opponent to a desirable position and in the process, we may lose the flow of the move that is the essence of Aikido. Therefore, it is advisable to prepare your stance before you are tightly grabbed.

I always instruct in my class that as we train in a martial art, if we are to be grabbed, we should not allow the opponent to have 100% control of the hold or grab. It could be your preparatory stance to be able to move with higher degree of freedom, restrict the opponent's grab angle which will not maximize the strength of the hold, defend with your free arm to somewhat redirect the energy of the opponent, or even lower your center of gravity to weaken the body center of the opponent and his grab.

Let's remember that counter attacks while in motion are much more defensible than from a static position. Even if grabbed tightly, as we will discuss in the third element, don't suddenly grab the hand of the opponent to execute a counter technique, as you will encounter only a powerful attack. It is important to keep room to flex/maneuver your shoulders to weaken the grab to be able to move to the next step.

Fixation of Contact Point

If we go back to the principles of leverage, the fixation of the opponent's grab in the shoulder becomes part of the "simple levers, fulcrums, and points of forces" principle of leverage, so rather than untangle, it is best to fix (solidify) the contact point and make use of it. Fixation of the contact point is only the beginning of Nage's technique.

It is important to create at least one fixed place and use it as a starting point to move other parts of the body. What is important then is how to fixate the fulcrum (pivot point) and keep the other parts of the body fluid. There are over 650 skeletal muscles in the body and while they have to move in sync, the technique has to be selective in which muscles to use for the leveraging and keep the rest of the muscles on alert but not stiff. We may use flexor and extensor muscles of the arms, some of the pectoral and dorsal muscles, and some other muscular forces but maintaining fluidity in action is essential.

While we are not beasts, it is like imagining a cheetah attacking an antelope. The key is that as we are using leverages, and we can focus on certain muscles to make it happen.

Use of Reactionary Force of the Opponent

Obviously, just grabbing the opponent's hand on your shoulder does not lead to Nikyo. The opponent's grip is firm and twisting it to your preferred position is not easy. In order to unlock the strong grip on the shoulder, you may have to shift your center by Tai Sabaki. The easier way to move the opponent is to use their strength to your advantage. A Tai Sabaki, hip rotation, a combination of shoulder push and pull, shoulder rotation, Hizanuki (unplugging the knee 膝抜き) generates a reactionary force by the opponent. This is to say that when you keep the foundation of the body (the lower half) stable even when doing Tai Sabaki, and you keep the muscles fluid in the upper half, the opponent will have difficulty maintaining his or her center stable. This is when you can take advantage of the opponent's center destabilization.

Usually beginners, when grabbed by the shoulder, will try to apply technique with only the muscular strength of the arm and shoulder stiffening the majority of the body muscles. The moves are clumsy and very ineffective.

Construction of Leverage Complexes

Trying to steal the center of the opponent can take many forms. But once you have been grabbed or attacked in some way, you will have to respond by good Tai Sabaki, reposition yourself in a better position, and apply Aikido techniques that use leverage. The leverage can be multiple leverages taking place simultaneously or in sequence. Some are based on the upper body and some on destabilization using the lower body as when you are about to break the balance of the opponent and bring the opponent to the ground zig-zagging as in Ikkyo technique.

Some work with linear moves that are straight forward undulation and some with a spiral effect. Some have push and pull techniques leading to Kokyu Nage. Or a Tenchi Nage which *spirals* up one arm while keeping the other very low and destabilizing the core of the opponent.

We have to take into account that a single-dimensional leverage may not be sufficient to destabilize the opponent and that we have to wisely position as Nage to apply multiple "triangular" leverage complexes moving in different directions. The key will be to unlock the core trunk of the Uke to be able to execute a technique. Thinking and moving in the horizontal plane, as well as in the vertical plane, creates a very dynamic move for Aikido.

All this sounds very complex and difficult. But if the focus of one's training is to take into account these four points to achieve leverage, the progress in Aikido skills should be fast. I have been training in Aikido for over 55 years but I lament that I did not categorize the important moves of Aikido until we were confronted with the dojo closure during Covid and I had to give classes on Zoom format. It was then that I started to add many perspectives to my Aikido instruction to add value to my Zoom classes.

Repeating the training with a partner is essential to progress but it is also important to think about the theory and use those hypotheses to apply to training so that you can see the light, and get a serendipitous inspiration.

18 The Importance of Proper Breathing

In **Kendo** (剣道), it is said that you know your opponent's breath at the tip of his sword directed at you. The timing of all techniques and exchanges is based on **Hyoshi** (拍子), the rhythm, beat, or tempo, and it is important to know the timing of the opponent's breath to know when to move. This "rhythm" refers to the breath. Let's quote again from 17th century sword master Miyamoto Musashi.

The rhythms of martial arts are varied. First know the right rhythms and understand the wrong rhythms, and discern the appropriate rhythms from among great and small and slow and fast rhythms. Know the rhythms of spatial relations, and know the rhythms of reversal. These matters are specialties of martial science. Unless you understand these rhythms of reversal, your martial artistry will not be reliable.

The way to win in the battle according to military science is to know the rhythms of specific opponents, and use rhythms that your opponents do not expect, producing formless rhythms from rhythms of wisdom.

— *The Earth Scroll. The Book of Five Rings*, by Miyamoto Musashi, translated by Thomas Cleary (Barnes and Noble), 1993.

In martial arts, such as Kendo, where you hide your breath, and Judo, where you try to make your opponent miss his breath, breathing and the timing of your techniques are crucial.

In Aikido, in spite of having terms such as **Kokyu Ho** (breathing method 呼吸法), **Kokyuryoku** (breathing power 呼吸力), and Kokyu Nage (breathing throw 呼吸投げ), we do not have any specific training method for breathing. The instructions are rather abstract. Aikido takes breathing as an integral part of total body movement and while the explanation is simple and correct, it is difficult to implement it. Let's take a close look at the explanation of Ueshiba Moriteru Doshu, who explains breathing as follows:

In Aikido, the power to concentrate efficiently and exert oneself through natural movement according to the structure of the whole body is called "breathing power (Kokyuryoku 呼吸力*)." If you are caught up in the word "breathing (kokyu* 呼吸*)," you may think of "breathing" as "breathing matching" to overlap/match inhalation and exhalation, and of the timing, movement with the opponent.*

For example, when holding a teacup and drinking tea, no one drinks with shoulder or elbow's full strength. They bring the teacup up to their front and quickly bring it to their mouth. When carrying an object placed below, you would bend your knees and lower your stance to lift it up. People would naturally perform the most efficient movements without thinking, without using unnecessary force, to move in accordance with the structure of the body!

Breathing power (Kokyuryoku 呼吸力*), likewise, is efficiently exerted from a relaxed, natural posture.*

— *Aikido, Keiko to Kokoro*, by Ueshiba Moriteru, p 72, published by Naigai Shuppansha, 2018.

Often truth is described in such simple terms. A wise man's maxim is short and to the point, very enlightening but very difficult to implement.

Let me also add the description of breathing by second Doshu, Ueshiba Kisshomaru in his classic book of Aikido.

Physiologically, breathing in and breathing out is called breathing (Kokyu 呼吸*), and the breathing method described here is exactly the same. However, it also means breathing with the entire body, from head to toe, with heaven and earth. In other words, in this* *case, breathing is the very act of becoming one with the flow of heaven and earth and being in harmony with nature. It is in cultivating such a state of mind that we can fully concentrate our natural power.*

— *Aikido*, by Ueshiba Kisshomaru, p 152, published by Kowado, 1957.

The descriptions by both Doshus are indeed very abstract. However, I read in both the very clear terms that for Aikido to be effective, you have to mobilize all your 206 to 213 bones, 250 to 350 skeletal joints, 650 to 840 skeletal muscles, about 900 ligaments, and approximately 4,000 tendons in an integrated manner in a "relaxed" and "balanced" motion. The flow of heaven and earth can only be achieved when the flow of Ki is in full operation. Stiffness, on the other hand, obstructs the natural flow of energy and Aikido cannot be perfected that way.

This abstract description of breathing power gives us a goal but not enough of the tools to understand the physiological aspects. Let's take a look at various aspects that may provide some hints.

When I was in my teens and in college, I got involved in many altercations and street fights. It is not that I was combative per se but somehow the society in Japan was very tense. As I was raised in South America, I was friendly and my eyes wandered around looking at people, but that caused major reactions from people that got offended by my eye stares intruding into their eyesight.

In Japan, that is called **Gan wo tobasu** (for lack of better translation, let's say "make a challenging eye contact" 眼を飛ばす). Kanai Sensei also told us stories of him getting in many street fights even when he was traveling as **Otomo** (traveling assistant お供) of O Sensei. Reading the biography of Tempu Nakamura, a famous martial artist and founder of Japanese Yoga who also became a teacher for numerous Aikido instructors such as Tada Hiroshi Sensei, Tohei Koichi Sensei from Shinshin Toitsu Aikido, and Sasaki Masando

Sensei, it shows he has an illustrious background of street fighting.

The reason that I am raising the street fight scene is that once you are about to enter into a fight, the tension pushes you to hyperventilate (shallow/sometimes deep and short breathing) and you get lots of symptoms of dry mouth/throat and body stiffness caused by the high tension. Of course, it was pre-Aikido training or in the very early stages of my Aikido training, so I can imagine that my fight or flight response training was still very limited. I am a peaceful person now, but the reason I am raising this issue is that I have learned that breathing had a lot to do with it. Perhaps the same can occur before a big music performance or an important job interview.

Breathing is much more than a physical and mechanical function to inhale air (O_2) and exhale CO_2. We need oxygen to survive but the body also needs some carbon dioxide to function. By hyperventilating, the oxygen that goes into the system to exchange with carbon dioxide may take out too much carbon dioxide from the body. It sounds good, right? The answer is that a suitable amount of carbon dioxide is necessary for the system to maintain a good chemical balance. Having the proper amount of carbon dioxide is one of the ways that the body relaxes. A proper amount of carbon dioxide is essential for the muscles to relax and function at their peak. By hyperventilating and losing carbon dioxide in the system, muscles contract.

Even though it is a bit technical, allow me to post a portion of an article to which I contributed to a book about breathing.

Hypocapnia (a CO_2 deficit in the blood and other body fluids, a deregulated respiratory chemistry) is the result of overbreathing behavior, the mismatch of breathing rate and depth. Its consequence is an increased level of pH, or respiratory alkalosis, which may have profound immediate and long-term effects that trigger, exacerbate, and/or cause a wide variety of emotional, perceptual, cognitive, attention, behavioral, and physical deficits that may seriously impact *health and performance. Although the fundamental importance of CO_2 in body chemistry regulation is common knowledge to any pulmonary or acid-base physiologist, it remains virtually unknown by most healthcare practitioners, health educators, breathing trainers, and laypeople.*

Here are some of the acute (immediate) symptoms triggered by overbreathing (Hypocapnia):

BREATHING: shortness of breath, breathlessness, feelings of suffocation
CHEST: chest tightness, pressure, and pain
SKIN: sweaty, cold, tingling, and numbness
HEART: palpitations, irregularities, rate
EMOTION: anxiety, apprehension, worry, emotional outburst, crying
STRESS: tenseness, fatigue, weakness, headache, hypertension
HEAD: dizziness, lightheadedness, fainting, black-out, confusion, disorientation
SENSES: blurred vision, dry mouth, sound seems distant, reduced pain threshold
PSYCHOLOGICAL SHIFTS: memory discharge, hallucinations, self-esteem
COGNITION: attention deficit, loss of focus, inability to think
CONSCIOUSNESS: feelings of "other worldliness," sense of disconnectedness
PERIPHERAL CHANGES: trembling, twitching, and shivering
MUSCLES: tetany, spasm, weakness, fatigue, and pain
— *Good Breathing, Bad Breathing*, by P.M Litchfield, Izawa Kei, Tsuda Akira, published by "Gendai no Espri" Shibundo, 2006.

Now that I have explained about the negative impact of imbalanced breathing, let's go back and use the criteria of other Japanese Budo when they explain about breathing. I am in total agreement with the comments of both Doshus, but I'd like to add more specifics to reach the natural way to breathe while executing Aikido techniques. The following samples are from Karate experts.

- Abdominal breathing is a breathing technique in which the mind is focused on the Tanden, the lower abdomen is inflated, and the diaphragm is lowered, allowing air to enter the lungs.
- The difference between abdominal breathing (**Fukushiki Kokyu** 腹式呼吸) and thoracic breathing (**Kyoushiki** 胸式呼吸) is whether the shoulders remain still or move. This difference is very important. Naturally, with abdominal breathing, there is no vertical movement of the shoulders. If the shoulders move up and down, the opponent will be able to read the beat and take advantage of it.
- Breathing in should go to the maximum and be limited to 70% to 80% of lung capacity, with 30% left after exhalation, not a full chest nor empty lungs. If you exhale fully, you will not have any power and your movements will be breathless, which is very unfavorable. Inhaling a full chest of air will cause the body to slow down and become less responsive. It is important to train and be mindful of your breathing so that you can turn to exhale at any time when you are inhaling. If you are attacked when you are breathing in with your chest full, you will not be able to turn to exhale, and your breath will stop for a moment, and your movement will also stop.
- When we become impatient, our breathing becomes disturbed, and we unknowingly start to breathe through our chest. When we breathe with our chest, our Ki (気) center, which has been stored stably in the Tanden, gradually rises higher. As the Ki center gradually rises to the top, feelings of agitation arise. When the Ki center rises to the top of the head, you are not grounded, you are in a state of panic, your surroundings are blank, and you have no

idea what is going on. When you are slumped over in a Kumite match, you are usually in a state where you have forgotten how to breathe because your mind is so pumped up.

- When you are fighting against someone of a higher rank than yourself, even though you are not doing any techniques or strenuous movements, your breathing becomes erratic, you sweat profusely, and you may feel very tired. In other words, it is in such a moment that you should do abdominal breathing to calm the mind, nurture equanimity, and lead yourself to a state of nothingness (**Mu** 無).
- Karate experts advise that breathing while in Kamae (preparedness stance 構え) posture should be quiet and slow and flowing, with no breaks between inhalation and exhalation. If a person is attacked while inhaling, there will be a delay of one step in the counterattack. In particular, the moment when the body reacts most sluggishly is the moment when the chest is fully inhaled and just before the exhale; that is, the moment when the body turns from inhalation to exhalation. In order to avoid being attacked at this point, you must never allow your opponent to read your breathing.
- If you feel you are suffocating and struggling to breathe properly, most likely the Ki is stagnant due to tension, causing the body to become stiff and sluggish. In such a case, taking a wider Maai (distance with opponent 間合い) for safety, do breathe in, and exhaling slowly to improve the flow of Ki and relieve stiffness in the body. Remember to fix the hypocapnia situation.

It is easy to talk about breathing methods, abdominal breathing, and so forth. When it comes to actual practice, among the ones I recommend are the following two.

One is by American celebrity Dr. Andrew Weil, who advocates for alternative medicine including the 4-7-8 breathing technique. He recommends three ways to breathe, but I personally like the 4-7-8 breathing technique. It goes as follows:

The 4-7-8 breathing exercise is utterly simple, takes almost no time, requires no equipment and can be done anywhere. Although you can do it in any position, sit with your back straight while learning the exercise. Place the tip of your tongue against the ridge of tissue just behind your upper front teeth, and keep it there through the entire exercise. You will be exhaling through your mouth around your tongue; try pursing your lips slightly if this seems awkward.

- Exhale completely through your mouth, making a whoosh sound.
- Close your mouth and inhale quietly through your nose to a mental count of **four**.
- Hold your breath for a count of **seven**.
- Exhale completely through your mouth, making a whoosh sound to a count of **eight**.
- This is one breath cycle. Now inhale again and repeat the cycle three more times for a total of four cycles.

— Source: *https://www.drweil.com/health-wellness/body-mind-spirit/stress-anxiety/breathing-three-exercises/*

Another breathing technique I like is humming while breathing through your nose. Humming causes the air to oscillate, which in turn greatly speeds up the exchange of air between the sinuses and the nasal cavity. Nitric oxide produced through nasal breathing, especially if humming, not only improves heart function, blood pressure, immunity, and cognitive function, decreases inflammation, but also helps improve your performance during a workout and decreases anxiety and depression. Nitric oxide helps cells throughout the body destroy pathogens, and it's also both a vasodilator (it helps blood vessels open wider) and a bronchodilator (it expands airways). It sure looks like

having more Nitric oxide in the system would help the flow of Ki, allow more relaxation and better flexibility.
— Source: *Nasal Nitric Oxide: Can you HUM your way to better health? https://www.youtube.com/watch?v=6v-nTRLEXFk*

Now let's go back to the Aikido definitions of breathing power and what we can learn from science and other martial arts to achieve the descriptions by both Doshus. We have understood that we can regulate our breathing to make sure we do not stagnate the Ki flow. You can go every day to the dojo to train, but if you are not regulating your breath during normal hours, you will not be able to suddenly find magic while training in Aikido. You have to train in your balanced breathing at home, off the mat, find your optimum breathing on a daily regime that allows us to absorb Aikido flow more rapidly when we go to the dojo. So, improving our own daily lifestyle with proper breathing is imperative as part of Aikido training. Martial arts requires that your brain is in its most alert condition and your muscles can react the fastest and nimblest. We know it is not only through proper breathing, but if you wish to reach the level of fluidity that Ueshiba Moriteru Doshu describes, proper state of mind and good breathing definitely have to be part of it.

On the Aikido energy-based argument, one could say that breathing methods can be categorized as follows: **Gaiki Kokyu-ho** (外気呼吸法), which focuses on projecting the inner saved energy outwards, and the **Naiki Kokyu-ho** (内気呼吸法), which focuses in absorbing the abundant Ki energy existing in the universe into your body to energize your inner self. In the Aikikai system of training, we do not make much use of expressive **Kiai** (気合) but it does exist. Many other martial arts use Kiai. Then what is Kiai? Kiai is when you suddenly turn the tide and use the internally saved energy by Naiki Kokyu-ho to create a tsunami of outward bound energy using Gaiki Kokyu-ho. You could say that Kiai is also part of breathing technique.

For people in India and China, breathing has meant much more and has been an integral part of life and philosophical thinking. The Chinese letter for breath (息) is a combination of two letters, one meaning "self" (自) and the other meaning "heart/mind" (心). The Latin root word for breathing is "spir," which is more focused on the transaction of elements.

19 Reiho: Why Manners Matter

Budo (the Martial Way 武道) is the "martial" culture of Japan, which is a continuation of the warfare techniques known as **Bugei** (martial arts 武芸) and **Bujutsu** (martial techniques 武術). It is not merely a system of fighting "techniques" but has developed as a physical culture that includes a certain system of philosophy called "**Do**" (the Way 道).

Traditionally, senior martial artists have refused to add Budo in the category/genre of "sports," but as many of the Japanese Budo started to participate in the realm of international sports and the Japanese government set up a Sports Agency, Japanese martial arts were somewhat encouraged to accept classification/designation of "sports." The Sports Agency is one of Japan's administrative agencies. It is an external bureau of the Ministry of Education, Culture, Sports, Science and Technology, established on October 1, 2015, for the purpose of comprehensively promoting sports and other measures related to sports. Our Aikido is designated as a "demonstration sport" by the Agency. By the way, there is no precise translation of the word "sports" in Japanese.

Besides the mainstream Judo, Kendo, Karate, Sumo, Kyudo, Naginata, Shorinji Kempo, Jukendo, and Aikido that constitute the core of the Budo Council of Japan, there are over 90 martial arts historical groups categorized as **Kobudo** (Ancient Budo 古武道) that form an organization named **Japan Kobudo Promotion Society** (日本古武道振興会), rich in its martial tradition.

The topic of this chapter is **Reiho** (manners, etiquette methodology 礼法). And we will try to view how much of Reiho is connected to Budo, the Martial Way. Although Reiho on the surface looks just like a manners formality procedure, it has a deeper rationale.

The Samurai always had to assume an unexpected attack, and they learned the etiquette accompanied by defensive and counter-attacking stances. By practicing resourcefulness, working on the spur of the moment, and performing serious training, one could improve one's martial skills and acquire ultimate strength.

As you learned in Chapter 7, the etiquette of the Samurai was modeled on the Ogasawara school of etiquette that had been practiced since the Kamakura period (1185-1333), and was sublimated into a unique form of archery and sword etiquette by incorporating the logic of the warrior's spirit into it. While maintaining the respect and dignity inherent in etiquette, it is built around the movements of self-defense against the enemy and is a path that leads to the teaching of "keep it in the scabbard," where victory is won without pulling your sword out.

The "Reiho" conveyed by the Ogasawara school needs to be clearly distinguished from the way of "manners" as understood by modern people. Ogasawara ryu etiquette was the "behavior guideline of the Samurai" from the Kamakura period (1185-1333) to the Edo period (1603-1868). The basis of the art is to be aware of correct posture in the basic movements of standing, sitting, walking, bowing, holding things, and spinning, and to have a body that can support the neck and hips firmly. In other words, this traditional "Reiho etiquette" has an aspect of physical training that is necessary for a warrior to maintain or train his functions as a warrior.

For the Ogasawara school, "etiquette" is not just bowing before and after a match or as an etiquette greeting. The concept of "rite" is understood to be an aesthetic sense of "courtesy" that is incorporated into the daily gestures of the people, while assuming that the situation is wartime. The concept is succinctly expressed by the key words "practicality, simplicity, and beauty."

Samurai Reiho etiquette is a martial arts exercise that is similar in form to manners/etiquette classes and general etiquette, but completely different in awareness and breathing techniques. Iaido, in particular, is close quarters fighting for one's life! It requires the ability to respond to sudden situations. Therefore, every move you make is a technique. The philosophy of etiquette and the preparation of body posture and attitude in the movement against the enemy, the sense of tension and urgency, and the etiquette exercises to the point of exhaustion will cultivate a variety of awareness and abilities. It also leads to the strengthening of the martial arts, and strengthens the mind and body through etiquette alone.

We will take a look at Iai/Kendo, which has five major Kamae (postures 構え):

1. Chuudan no Kamae (Seigan no Kamae)
2. Jodan no Kamae
3. Gedan no Kamae
4. Waki Gamae
5. Hasso no Kamae

But Kendo people add two. which are not with a sword. One is Ashi Gamae (foot stance 足構え) and the other one is Kokoro Gamae (preparedness/attitude 心構え). There is another way to say Kokoro Gamae and that is Ki Gamae (energetic stance 気構え), which may sound more appealing to Aikido people. Kokoro Gamae or Ki Gamae are an attitude and aura that overflows from the person who is well prepared.

In Ki Gamae, the person's posture is so relaxed but has no openings nor shows any hint of weakness. This is what I think was intended by the original teachings of Reiho by the Ogasawara school. By maintaining a well-defined stance, you can intimidate your opponent so that he or she cannot strike you. If you feel that your opponent is strong when you are in a stance with him or her, it can be said that his or her stance is well-developed. Improve your Reiho through posture and acquire the skill to intimidate your opponent.

Let's talk about the background of Reiho using the case of "one beat." In Bujutsu (martial techniques 武術), the basic principle is to perform as many sequential movements in one beat (**Ippaku** 一拍 or **Hyoshi** 拍子) as possible to prevent the enemy from reading your actions. Most martial arts techniques are performed in one beat as much as possible. Here are two examples from Iaido.

"Kneel down at the same time you put your hand on the sword, prepare to draw one's sword (**Koiguchi wo Kiru** 鯉口を切る), step out with one foot, and draw the sword and cut." In this way, a technique can only be realized when it encompasses a myriad of actions all at once.

"Putting one's hand on the sword. Kneeling down. Next, one stands on both knees. Then prepare to draw one's sword (Koiguchi wo Kiru 鯉口を切る). Then one puts one's leg out in front of the other in the iai-koshi position. Then one draws the sword. And finally cut."

If the movements were divided in this way, it would be easier to learn and perform, but it also makes the technique useless because the sequential moves are so fragmented. Advanced swordsmen may intentionally divide the movements to mislead the opponent by using them as feints, but they are only for advanced applications. In reality, the enemy in front of you does not wait for you to move, and martial arts do not like movements with multiple beats.

In the case of etiquette, however, the opposite is true. It is a habit and custom that comes from the knowledge of the ever-present battlefield, but it would be extremely impolite to show wariness, harmful intent, or hostility toward one's opponents in times of peace.

Therefore, it is necessary to be aware of where the sword is placed, how it is placed, how one carries oneself, and even one's every move. This awareness is the very spirit that forms the foundation of etiquette, and it is called **Kikubari** (attentiveness 気配り) and **Kokoro Kubari** (mindfulness 心配り).

Even so, if we change according to the circumstances of each situation, different people will perceive it in different ways, and it is possible that no matter how much we try to be attentive, we may still cause others to feel that they have been treated disrespectfully. Therefore, the Samurai warrior etiquette embodied the spirit of attentiveness and established certain rules to be shared.

In a Samurai family that carries weapons, "one beat (Ippaku 一拍)," a single beat of action, can be difficult to detect and extremely unpleasant. A sword held to the left, right, or back, or a small dagger at the waist may be drawn in an instant. In the same way, in etiquette, "Ippaku" is prohibited in principle, as it can cause alarm or at least great discomfort.

In Iaido, in particular, the Samurai may locate the sword or the direction of the blade in a way that shows one has no hidden intention to suddenly attack in one breath. For example, as most people were right-handed, while sitting, they would place the katana on the right side so that it takes an additional step to take the sword out of the scabbard.

The Reiho can go beyond the attentiveness to others. It is also a matter of how to be prepared at all times. As the famous 17th-century swordmaster, Miyamoto Musashi, wrote:

"In the science of martial arts, the state of mind should remain the same as normal. In ordinary circumstances as well as practicing martial arts, let there be no change at all—with the mind open and

direct, neither tense nor lax, centering the mind so that there is no imbalance, calmly relax your mind, and savor this moment of ease thoroughly so that the relaxation does not stop its relaxation for even an instant."

— *The Water Scroll. The Book of Five Rings*, by Miyamoto Musashi, translated by Thomas Cleary (Barnes and Noble), 1993.

The ideal of a Samurai was to perform well on the battlefield, and it was only on the battlefield that he was honored to die. Therefore, peacetime was for preparation for war, and the mindset of peacetime was to always be ready for battle. If we translate this into the fighting posture, it is the state of "readiness" to attack and defend at any time.

In the Japanese art of war, martial arts were used in times of war, while etiquette was used in times of peace. Therefore, etiquette is a Kamae (preparedness stance 構え) posture.

What is needed in peacetime is a defensive stance, not an offensive one, and what is needed in response to an attack in peacetime is an expression of respect for others. Therefore, the principles of the martial arts can be utilized in peacetime. Without knowing the martial arts, it is impossible to speak of Samurai etiquette. Here again, it is clear that the mere rituals of Samurai etiquette are not the essence of Samurai etiquette.

Conversely, the essence of the etiquette of the Samurai family does not originate in the rituals of the Imperial court, but in the martial arts, which are involved in the exchange of life. Therefore, it is not formal, but it has an effective and physical rationality. Because it is not a ritual filled with mythical meanings, but a gesture grounded in gravity and physical structure, it transcends time and culture and is valid for modern people as well.

Posture of Kamae

We have covered a bit in Chapter 5 when we discussed posture but let me repeat to make a case from the Reiho perspective. The posture of Kamae is not forceful and not relaxed. This is because neither tension nor relaxation can be changed instantaneously; i.e., they are in a state of "staying" (stagnation). The basis of "stance," which is the maintenance of optimum tension, is to keep the joints slightly bent without fully extending them (not consciously bending them). A posture with fully extended joints is a posture in which no further movement is possible, and it is full of gaps. Naturally, the basic principle of martial arts is to avoid showing any gaps.

The legs are shoulder-width apart, facing in the direction of movement (not outward), knees lightly bent, hips relaxed, and hip joints also lightly loosened. In this way, the upper body from the waist up naturally becomes vertical (without stretching the chest). Since the warrior wears hakama, even if the hip joints and knees are slightly bent, it does not look as if they are bent from the outside and is not unsightly.

The elbows are also lightly bent and the arms are lowered to a natural position, neither directly to the side nor directly in front of the body, and the hands are placed on the tops of the thighs.

The third joints of the four fingers of each hand, except the thumbs, should be lightly bent, and the backs of the hands should be in a low, mountainous shape, with the thumbs resting against their sides.

Avoiding Rough Mistakes

There is nothing more disgraceful than a Samurai, preparing to go to the battlefield, who falls down and injures himself at home before the battle. Therefore, safety (prudence) is the top priority in the behavior of a Samurai at home. In Samurai etiquette, avoiding mistakes is the top priority, even more so than courtesy. Interestingly, the etiquette was formulated to avoid tripping and slipping, to avoid twisting and other joint-intensive movements, to avoid dropping objects, and to pursue the proper way to walk, sit, stand, change direction, and hold objects (these are practically taught in modern etiquette classes). The pursuit of rationality of movement based on safety is the reality of Samurai etiquette.

If you are a frequent traveler to Japan, you may have likely observed the checking procedures that train conductors perform while conducting the trains or before the train departs. Not just visually checking but pointing fingers of the targeted points or repeating verbally the confirmation action they are taking.

This action is called **Shisa Kanko** (pointing and calling 指差喚呼). While train conductors are not martial artists, it is obvious that the Japanese railroad system has incorporated the concept of Reiho principles. You can see from this example how Reiho can be a wonderful defensive procedure as well.

Reiho is a very deep concept ingrained in Japanese martial arts. It is not a religious ritual as some wish to highlight but has fundamentally a more human element to it.

Kanai Mitsunari Sensei from Boston summarized it elegantly as follows:

At any rate, people working in martial arts tend to become attached to technical strength. They become arrogant and boorish, bragging of their accomplishments. They tend to make unpolished statements based on egoism. They immerse themselves in self-satisfaction. They not only fail to contribute anything to society but, as human beings, their attitudes are under-developed and their actions are childish. What is important about Reigi-saho is that it is not simply a matter of bowing properly. The basis of Reigi-saho is the accomplishment of the purified inner self and the personal dignity essential to the martial artist.

If we advance this way of thinking, the matter of Reigi-saho becomes the question of how one should live life itself. It determines what one's mental frame and physical posture should be prior to any conflict situation; the guard-posture must have no openings. Thus, Reigi-saho originates in a sincere and serious confrontation with life and death. Above all, Reigi-saho is an expression of mutual respect in person-to-person encounters, a respect for each other's personalities, a respect which results from the martial artist's confrontations with life-or-death situations. The culmination of the martial artist's experience is the expression of love for all of humanity. This expression of love for all of humanity is Reigi-saho.

To read more of Kanai Sensei's philosophy of Reiho, see his essay on the topic:
http://www.neaikikai.org/writing-and-memorials/2016/3/12/s7q79alqavzwgenak0eagi14p601io

20 Beyond Martialness, Healthy Benefits

Many books have been written on Aikido. The topics range from techniques, training, education, gender, philosophies, applications to conflict resolution, business, history, and many more. In reviewing many of the books on Aikido, one finds a very broad categorization of its benefits as self-defense, stamina improvement, stress relief, business negotiations, and so many more topics. Interestingly, Aikido is not viewed solely in the context of martial arts but covers such a wide spectrum that you may call it Aikido lifestyle. While many of those books do not highlight it directly, many Aikido practitioners have branched out or got cross-trained in the field of health.

I know that many martial and combat training programs are touted as beneficial for weight loss or cardiovascular exercises and I do not doubt that they have a positive impact on many health benefits, but Aikido seems to be associated with a much broader sense of health associated benefits. Examples abound, and I will highlight a few of those.

Aikido is indirectly or directly connected to diet practice of Macrobiotic, stretch exercises of Makko-ho (真向法), circulation and body realignment exercises of **Nishi Katsuzo** (西勝造) system, breathing and body/mind exercises of **Nakamura Tempu** (中村天風), **Jikyou Jutsu** (a Japanese method of health exercises 自彊術), and the list goes on.

The Aikido founder may not have invented those health initiatives but many of his disciples incorporated it into the daily Aikido training and lifestyle. For example, as described in Chapter 9, Makko-ho (真向法)—four major stretch moves—are incorporated as part of the warm-up exercises from the time of Kisshomaru Doshu and are passed on to current Doshu and Dojocho.

Epilogue

When I was studying at Keio University, I read a book in my cultural anthropological studies that rice-cultivating societies like Japan, compared to many hunting-based societies, focused more on copying from those farmers that had the highest yield, best farming practices, or best tasting produce. The communication did not have to be very precise, and greetings always involved elements of empathy and gratitude in the communal work environment.

I still have difficulty in translating words like **Yoroshiku Onegaishimasu** (something along the lines of "please treat me well"), **Otsukaresama** (something along the lines of "you must be tired as you had a rough day today"), **Gokurosama** (something along the lines of "you went through a lot of suffering"), **Okagesama** (something along the lines of "thanking you for your backing"), **Osewa ni Narimashita** (something along the lines of "appreciate all you have done to me"), and when you leave work before others you may say **Osaki ni Shitsurei Shimasu** (something along the lines of "I am not supposed to… but pardon me for leaving before you"). Imagine the last one in an office environment when you leave work ahead of the others that are doing overtime work, often without pay. You are entitled to go but the community obligations are not met.

I was brought up in South America until I finished elementary school. Upon returning to Japan and hearing these terms, it caused me to wonder why I had to be so thankful to others when the help was minimal. However, in the aftermath of the Kobe earthquake in 1995 and the Tohoku earthquake in 2011, witnessing a self-supporting social order in those devastated communities and how people helped each other, I understood and was deeply moved.

I believe Budo instruction and training in Japan also has some similarities with this farming communal tradition. We start with

Yoroshiku Onegaishimasu and finish with Arigato Gozaimashita, or we could use **Otsukaresama Deshita** (it was hard training, you must be tired) as well. But most important for all is that training instruction is the same in trying to copy from the teacher.

It is up to the student to observe carefully and try to copy exactly as the teacher demonstrated. While in today's Japan, we see a more Western approach of providing clearer instructions to teach techniques, traditionally you were expected to learn it by actual personal experience without many verbal explanations. It was like an apprenticeship. Your dedication to the art, your focused attention to detail, your repetitive efforts to improve, and your loyalty to your teacher were the key to this system. It was acceptable to steal the know-how from your teacher.

Ueshiba Kisshomaru Doshu writes in the biography of O Sensei as follows:

O Sensei focused his teaching on the basics; what he demonstrated for students was always clear and simple. When I had questions about how the hands or feet should move in a particular technique, he would just throw me, and then say with a big smile, "That's how it's done—did you get it?" I believe he felt that technique could not be taught in words; he wanted people to learn by being thrown over and over again. Each time the same technique is used, it differs, depending on the angle, the timing, or the force applied by each partner—these crucial variations may be quite subtle. I believe O Sensei saw little value in words because they fixed the form of a technique without capturing its fluidity in actual practice.

— *A Life in Aikido*, by Ueshiba Kisshomaru, translated by Izawa Kei, Mary Fuller, published by Kodansha, 2008.

I have lived more than two thirds of my life outside of Japan. When I worked for the automotive giant General Motors, I was involved as a team member in joint venture negotiations with Japanese automotive firms like Toyota, Suzuki, and Isuzu. I believe I contributed to bridging the cultural gap that existed between two quite different business cultures. With this business background training, I have tried to do similar bridging in Aikido with those overseas people interested in it, as most Japanese instructors do not explain the background of how things are done in Japan.

I totally agree with Kisshomaru Doshu and current Moriteru Doshu that there is nothing better than you can personally experience applying some techniques and get the bodily sensation of making it happen. But I also believe some little footnotes, not necessarily theories, are useful for people who are not acquainted with Japanese culture. I also tried to add my personal observations of Kanai Hambei Mitsunari Sensei's Aikido teaching, and I hope with the collaboration of my kohai and lifelong friend Itoh Yasumasa Sensei from Tekkojuku Boston, we can sometime soon we can catalog some of Kanai Sensei's techniques in some visual form.

I was very fortunate to have met many direct disciples of the founder of Aikido over my course of over 55 years of Aikido training. Due to my professional job requirements, I have lived in many countries and as such also met with many very inquisitive and high level Aikidokas. One of them was Kanai Hambei Mitsunari Shihan who resided in Boston and was the founder of New England Aikikai, where I trained for two years in the mid to late '70s and established a life-long teacher-student bond with him. Kanai Sensei was Uchi Deshi (live-in disciple) at the Ueshiba family, contemporary of Tamura Sensei, Chiba Sensei, Kurita Sensei, and some others. He was one of the last Uchi Deshis under O Sensei and Kisshomaru 2nd Doshu.

Kanai Sensei built a reputation of having a very dynamic and perhaps a very esthetical form of flowing Aikido. His range of movements were very wide and the result was that his projections were very powerful as well. Kanai Sensei also studied the potential applications of the Aikido moves and as such he created many moves that are admired even today. His signature hair wave as he did his techniques aside, his posture and style were impeccable.

Kanai Sensei studied physiology on his own and would study and consult with the *Gray's Anatomy* book which was prominently located in his bookshelf. He surprised many of those who were in the physical therapy or health related business with his profound knowledge of human anatomy. He was a very Japanese traditional individual with a detailed approach to everything. He studied metallurgy as well and even made his own tsubas (sword guards) and tantos (daggers). He was a detailed craftsman and he also made his own Menukis, Fuchi Kashiras, and many sword ornamental items. His knowledge of swords was so advanced that he helped the Boston Museum Japanese Swords collection curators.

What impressed me the most was that all these were not separate isolated interests. For Kanai Sensei, it meant that all were part of a much bigger picture. His anatomy knowledge was interconnected to the metallurgy of the swords, the sword tsuka and the shape of the sword for a dynamic and projective slicing movement. Even the sword ornaments were part of this expanded world. Kanai Sensei would tell us now and then how he felt when he attacked O Sensei and the feeling of being thrown as he bounced away from O Sensei. He was determined to decipher some of the powerful projections executed by O Sensei. From the bouncing experience, Kanai Sensei developed some techniques he called "Hajiki Goshi" (Hajiku is to bounce and Goshi is from Koshi which means hip). Kanai Sensei added names to the various moves he developed which are not well known outside his direct students or those who followed him.

While Kanai Sensei had many spiritual aspects, I should say that he was more of an Aikido mechanical engineer, perhaps more a tear-down analyst of O Sensei moves. I never had the chance to see O Sensei alive as I started Aikido at Hombu dojo a couple of weeks after O Sensei passed away, but from all I hear and I read, I conclude that O Sensei was more a natural genius, not bound in stifling structures even though he had his details that were also very accurate and correct. Aikido flowed naturally for O Sensei and while perhaps his mastery of the moves were based on the hard training and dedication, one cannot forget that he had a genius in him.

Kanai Sensei's approach to things, based on his broad knowledge, was to find or replicate the basis of that energy that O Sensei generated when he would throw Kanai Sensei. O Sensei didn't teach details and young Kanai Sensei, as lost as he was, gradually started to build his own interpretation of O Sensei's mechanical moves. He tried to understand O Sensei's moves but at the same time, he tried to strengthen the martialness of Aikido. For example, his Aikido Hanmi standing stance is not the same as many Aikidokas, as he took a stance similar to that of Kendo or Iai, rather than the way most people do it today in Aikido.

Kanai Sensei emphasized the consolidation of the Nage's power when executing a technique. For him, the directional stance of the energy projection had to be unified. He called this stance Toitsutai (統一体), which means basically "integrated body." What he meant by it was to align the directions of the eyes, shoulders, arms, hip, and the feet in the direction one was projecting the energy.

For Kanai Sensei, he tried to understand the power of Aikido in very mechanical terms. He had a very rational mind for Aikido, his tear-down analyst aspect of viewing Aikido was very strong and obvious.

The interest in technical matters I got by training under Kanai Sensei never went away. I continued to visit him in Cambridge almost yearly and encouraged him to write a book on his Aikido technical perspective, which to me was very unique. We started to communicate via fax and I helped him translate some of the texts of his rather complex explanations about the philosophy and his technical interpretations of Aikido. This was in the early ’90s when emails were about to become available. Kanai Sensei, being a very thorough individual, wrote in traditional Japanese writing style, which was rather difficult to translate into English and I think a lot was lost in translation. However, this effort gave fruition to a video called “Technical Aikido.”

Kanai Sensei passed away back in 2004 and now it has become my turn to take a close look at some of his teachings so that whatever I know I could pass it on. This book focused more on the mechanical functions of the different parts of the body in Aikido execution as explained by Kanai Sensei but not enough on the analysis of the techniques themselves. As I wrote “in the beginning,” I tried to add many cultural background explanations of Japan related to martial arts. Explaining the footwear and walking methods of the historic Japanese may not sound as connected to today’s Aikido, but to me, all these details seem to have a tremendous connection to the formation of Aikido. Aikido was born within these cultural parameters, just as the Afro-Brazilian martial art of Capoeira is greatly affected by the elements of dance, acrobatics, music, and spirituality of Brazil.

This book does not ignore the great achievements of O Sensei, but rather, it attempts to provide a complementary explanation of the unique Japanese cultural context and physical functions in Aikido that the many books written on Aikido have not covered.

Acknowledgments

Looking back at my professional life, I see that I spent a substantial amount of my energy as a cultural interpreter, even when I was working for the government or working for General Motors on multi-billion dollar projects. Working for the International Aikido Federation also required a lot of delicate intra-cultural diplomacy. Having lived and worked professionally in four continents gave me the preparation stage to sharpen my tools. *Balance in Motion* is a small attempt to extend the cultural interpretation surrounding Aikido. I hope it is a good starting point for many more people going deeper into this interesting genre.

Books can only be written with interaction with other people. I have dedicated this book to Kanai Mitsunari Sensei and my wife Mariquita. To Kanai Sensei because he was instrumental in decompartmentalizing Aikido moves and combining/transforming them into techniques not seen in many other mainline instructors. His Aikido had value-added teachings and forced those who studied under him to go outside the box to find the limitless applications of body movements.

To my wife Mariquita, not only for her illustrations found throughout this book, but for her sincere and constructive criticism of my Aikido movements. As a creative artist combined with her long experience in Aikido at Hombu Dojo and her encounter with Kanai Sensei, Mariquita acquired good eyes to analyze the art of Aikido.

I need to acknowledge the significant contribution of Itoh Yasumasa Sensei, Chief Instructor of Tekkojuku in Boston, with whom I have had countless discussions on Kanai Sensei's Aikido and his movements. Itoh Sensei was close to Kanai Sensei for 25 years and was instrumental in working with Kanai Sensei to classify and name Kanai-style Aikido techniques. The classification of the names gives a great understanding of Kanai Sensei's Aikido and his favorite body movements.

The interaction I had with Tanshinjuku advanced students Tim Cunningham, Jim Renwick, Kent Krumvieda, Reina Krumvieda Watanabe, and Alex Magaña during and after the COVID-19 pandemic time gave me a great opportunity to hypothesize and corroborate the concept of this book. The suggestion to write a book came from Richard Shouse, a former student who was in the medical profession. I would like to extend my gratitude to mentors, colleagues, and Aikido instructors who contributed to the author's understanding and development.

I also thank my Mother, Etsuko Izawa (1904-1997), for the calligraphy that appears on the cover.

Last but not least, I need to thank Kara L. Stewart Sensei for taking on the tremendous task of being my editor and proofreader. Her experience as a writer of the book *Fall Down Seven Times, Stand Up Eight: Life Lessons for Everyday Warriors (e.g., the rest of us)* helped make the book *Balance in Motion* much more readable. The book's front and back covers, interior layout, and design are the brilliant creation of another Aikido friend, Andrew Blevins Sensei, founder of Kiryu Aikido dojo. This beautiful book is a result of his vision and talent.

Glossary

A

Agura (slightly looser sitting style 胡座)
Aikido (the Way of Harmonizing Energy 合気道)
Aiki Otoshi (pulling on the opponent from the side, grabbing the back of Uke's knee from the front, lifting Uke up, and throwing Uke backwards 合気落とし)
Anza (slightly looser sitting style 安坐)
Ashi Gamae (foot stance 足構え)
Ashi-garami (joint technique 足緘)
Ashinaka (half-foot sandal 足半)
Ashi Sabaki (feet movement 足捌き)
Atemi (a blow/strike to prevent Uke from regaining attack stance 当身)
Atemi Waza (striking techniques including kicks 当身技)

B

Bokken (wooden sword 木剣)
Budo (the Martial Way 武道)
Bugei (martial arts 武芸)
Bujutsu (martial techniques 武術)

C

Chikara wo Nuku (release muscle tension 力を抜く)
Chutanden (middle Tanden 中丹田)

D

Daishoh (matched long/short pair of traditionally made Japanese swords 大小)
Deashi Barai (forward foot sweep 出足払い)

Do (the Way 道)
Do-Yuh (friend of the path 道友)

F

Fukugo Teko no Kouchiku (construction of leverage complexes 複合梃子の構築)
Fukushiki Kokyu (abdominal breathing 腹式呼吸)
Funakogi Undo (rowing exercise 船漕ぎ運動)

G

Gaiki Kokyu-ho (breathing method of the outside Ki substance that is the source of life energy that is ubiquitous in the natural world 外気呼吸法)
Gan wo Tobasu (make challenging eye contact 眼を飛ばす)
Geta (wooden clog with a strap between your big toe and the rest 下駄)

H

Hagai Jime (rear bear hug 羽交い締め)
Hajiki (bouncing 弾き)
Hakama (pleated pants 袴)
Handoryoku no Riyou (use of reactionary force of the opponent 反動力の利用)
Haniwa (clay figurines 埴輪)
Hanmi (half body 半身)
Happo-Kuzushi (collapsing the opponent in eight directions 八方崩し)
Hara (belly 腹)
Hikite (the left hand is called "Hikite" and mainly holds the sleeve of the opponent 引き手)
Hikyaku (messenger, literally meaning flying legs 飛脚)
Hinawaju (matchlock gun 火縄銃)

Hiza (knee 膝)
Hizanuki (unplugging the knee 膝抜き)
Hyoshi (rhythm, beat, tempo 拍子)

I

Ichigan Nisoku Santan Shiriki (priority of important points in training: First visual alertness, Second foot movements, Third courage, Fourth physical power 一眼二足三胆四力)
Ikioi (momentum, vigor, force 勢い)
Ikkyo (first teaching 一教)
Ippaku (one beat 一拍)
Irimi (entering, sliding in 入身)
Irimi Nage (entering throw 入身投げ)
Issoku itto no maai (distance of one step, one sword 一足一刀の間合い)

J

Japan Kobudo Promotion Society (日本古武道振興会)
Japanese Dance (日本舞踊)
Jikyou Jutsu (a Japanese method of health exercises 自彊術)
Jiten (pivot 自転)
Jo (staff 杖)
Juji-Garami (cross elbow interaction 十字絡み)
Jujutsu (柔術)

K

Kaeshi Waza (counter technique 返し技)
Kaiten (turning 回転)
Kamae (preparedness stance 構え)
Kamitanden (upper Tanden 上丹田)
Kamiza (dojo front, seat of honor 上座)
Kansetsu Waza (joint techniques 関節技)

Kata (strict forms 型)
Kata (shoulder(s) 肩)
Kata Hiki Otoshi (pull shoulder drop 肩引き落とし)
Katame Waza (restraining techniques 固め技)
Katana (sword 刀)
Katate Tori (one hand hold 片手取り)
Katatori (shoulder grab 肩取り)
Keiko Aite (training partner 稽古相手)
Keikogi (training outfit 稽古衣)
Kendo (剣道)
Kenkokotsu (scapula 肩甲骨)
Kiai (the momentum of the mind when one concentrates one's mind on a task; also, a call to action at that time 気合)
Ki Gamae (energetic stance 気構え)
Kikubari (attentiveness 気配り)
Kime no Kata (Judo kata 極の形)
Kimono (traditional clothing 着物)
Ki no Nagare (flow of Ki energy 気の流れ)
Kiryoku (power of ki 気力)
Kiza (kneeling posture with buttocks on heels 騎座)
Kobudo (Ancient Budo 古武道)
Kobujutsu (ancient martial techniques called Jujutsu and Taijutsu 古武術)
Kodokan (the headquarters of the worldwide Judo community 講道館)
Kohgake (foot instep guard 甲掛)
Koiguchi wo Kiru (prepare to draw one's sword 鯉口を切る)
Kokoro Gamae (preparedness/attitude 心構え)
Kokoro Kubari (mindfulness 心配り)
Kokyu Ho (breathing method 呼吸法)
Kokyu Nage (breath/breathing throw 呼吸投げ)
Kokyuryoku (breathing power 呼吸力)

Kosa Nage (leveraging against the opponent's elbow 交差投げ)
Koshi (hip to pelvis area; the kanji 腰 combines Nikuzuki [body flesh 月] and Kaname [pivot, pin, center, essence, substance 要])
Koshiita (lower back board 腰板)
Koshi ga Hairu (settled Koshi/tackling an issue seriously 腰が入る)
Koshi ga Hikui (low Koshi; humble 腰が低い)
Koshi ga Karui (light Koshi/quick to act 腰が軽い)
Koshi ga Omoi (heavy Koshi/slow to act 腰が重い)
Koshinage (hip throw 腰投げ)
Koshi Nuke (absent Koshi/coward 腰抜け)
Koshi Wari (Sumo squat 腰割り)
Kotegaeshi (flipping opponent's wrist 小手返し)
Kotodama (the spirit which is present in words 言霊)
Kubi ga Mawaranai (one cannot move the neck, meaning to be deep in debt 首が回らない)
Kubi Osae Kokyu Nage (neck holding Kokyu throw 首抑え呼吸投げ)
Kubi Otoshi (neck drop 首落とし)
Kubi Shime (neck choke 首絞め)
Kumite (grappling 組み手)
Kuzushi (destabilization 崩し)
Kyoushiki (thoracic breathing 胸式呼吸)
Kyudo (archery 弓道)

M

Maai (in-between distance 間合い)
Makikomi Otoshi (rolling/wrapping drop 巻き込み落とし)
Maki Otoshi (rolling drop 巻き落とし)
Makko-ho (stretch exercises 真向法 , currently used as warm-ups in Hombu Dojo)
Masakatsu Agatsu (True Victory is Self Victory 正勝吾勝)
Mata Wari (complete split 股割り)

Migi Hanmi (the right half of the body 右半身)
Misogi (purification 禊)
Morote Tori (two hand hold on one arm 諸手取り)
Mu (state of nothingness 無)

N
Nagare (flow 流れ)
Nage (technique initiator 投げ)
Nage Waza (throwing techniques 投技)
Naginata (halberd 薙刀)
Naiki Kokyu-ho (breathing method to take outside Ki substance into the body, transform it into inner Ki, and then change it qualitatively by increasing its density even more 内気呼吸法)
Nakamura Tempu (breathing and body/mind exercises by Tenpu Nakamura 中村天風)
Namba Aruki (a walking style that originated in the Edo period (1603-1868) ナンバ歩き)
Nenten (torsion or gyration 捻転)
Ne Waza (groundwork techniques 寝技)
Nihon Buyo (Japanese traditional dance 日本舞踊)
Nihonto (Japanese sword 日本刀)
Nishi Katsuzo (circulation and body realignment exercise system by **Katsuzo Nishi** 西勝造)
Nodowa (a thrust at the opponent's throat 喉輪)
Noh (能)
Nokogiri (鋸，のこぎり)

O
Obi (belt 帯)
Ogasawara school (小笠原流)
Osae Waza (pinning or suppression technique 抑え技)
Otomo (traveling assistant お供)

Oyobi Koshi (not fully engaged Koshi/lack of nerve or indecisiveness 及び腰)

R

Raku-za (worshipful legs or nobleman's seat; a way of sitting on the ground with the soles of both feet pressed together 楽坐)
Reiho (manners, etiquette methodology 礼法)
Rikisha (runners pulling light two-wheeled hooded vehicle 力車)
Rikko (scapula winging 立甲)
Ritsu-I (standing position 立位)
Ryo Kata Tori (both shoulder grab 両肩取り)
Ryote Tori (two hand hold 両手取り)

S

Sakkatsu-jutsu ("killing technique" 殺活術)
Sappou ("killing technique" 殺法)
Seika Tanden (sub-umbilical tanden 臍下丹田)
Seiza (squatting straight [also correct sitting] 正座 or 坐)
Senkotsu (sacrum bone 仙骨)
Sen no Sen (anticipate the attack or move of the opponent 先の先)
Sensei (teacher 先生)
Setten Kotei (fixation of contact point 接点固定)
Shikkoh (knee walking 膝行)
Shiko wo Fumu (Shiko stomping 四股を踏む)
Shime Waza (choking technique 絞め技)
Shimotanden (lower Tanden 下丹田)
Shinshin Toitsu (body and mind integration 心身統一)
Shisa Kanko (pointing and calling 指差喚呼)
Shisei (posture 姿勢)
Shitsuke (discipline 躾)
Shomen Uchi (front cutting attack to the head 正面打ち)
Sode Tori (sleeve grab 袖取り)

Sonkyo (bow and show respect; a sitting posture in which the body is rounded into a crouch or in which the knees are folded and the hips are lowered to a standing kneel 蹲踞)
Suburi (sword swing 素振り)
Sugata (appearance/shape 姿)
Sumi Otoshi (corner drop 隅落し)
Sumo (相撲)
Suriashi (sliding step すり足)
Suwari Waza or Zagi (sitting technique 座り技 or 座技)

T

Tachi or Bokken (wooden sword 太刀、木剣)
Tachiwaza (standing technique 立ち技)
Tai Sabaki (body movement 体捌き)
Tanden (body center 丹田)
Tanden ("Tan" medicine for long life 丹 and "Den" a place/field to make it 田)
Tanto (dagger 短刀)
Taoism (a Chinese philosophy 道教)
Tatehiza (raised knee sitting position 立膝)
Teko no Genri (principle of leveraging 梃子の原理)
Tenbin-Nage (leveraging against the opponent's elbow 天秤投げ)
Tenkai (turning 転回)
Tenkan (side backward turn 転換)
Tenshin (stepping back 転身)
Toitsutai (Integrated Body 統一体)
Tsuki (thrust 突き)
Tsuki Kotegaeshi (thrust to chest, flipping opponent's wrist 突き小手返し)
Tsurite (in the case of the right side body stance, the Tsurite hand is the right hand and mainly grips the collar 釣りて)
Tsuyo Goshi (strong Koshi/firm stance in business negotiations 強腰)

U

Uchideshi (live-in students 内弟子)
Ude Garami (joint technique 腕絨)
Ude Hishigi Hiza Gatame (joint technique 腕挫膝固)
Ude Hishigi Jujigatame (joint technique 腕挫十字固)
Ude Hishigi Udegatame (joint technique 腕挫腕固)
Uke (the attacker 受け)
Ukemis (rolls 受け身)
Ushiro Kubishime (rear hold and choke 後ろ首絞め)
Ushiro Ryo Kata Tori (both shoulder grab from back 後ろ両肩取り)

W

Wago no Budo (the art of harmony 和合の武道)
Waki wo Shimeru (tightening the armpits 脇を締める)
Wakizashi (short sword 脇差)
Waraji (straw sandals 草鞋)

Y

Yari (spear 槍)
Yabusame (horseback archery 流鏑馬)
Yokomen Uchi (strike neck and temple area 横面打ち)
Yo-tsu (low back pain 腰痛)
Yowa Goshi (weak Koshi/weak position in negotiations 弱腰)

Z

Zagi or Suwari-waza (sitting technique 座技 or 座り技)
Zagi Kokyu Ho (seated exercise 座技呼吸法)
Za-I (sitting position 座位)
Zanshin (unrelaxed alertness 残心)
Zarei (sitting bow 座礼)

Made in the USA
Monee, IL
04 November 2026

67374676R00105